EINSTEIN'S JEWISH SCIENCE

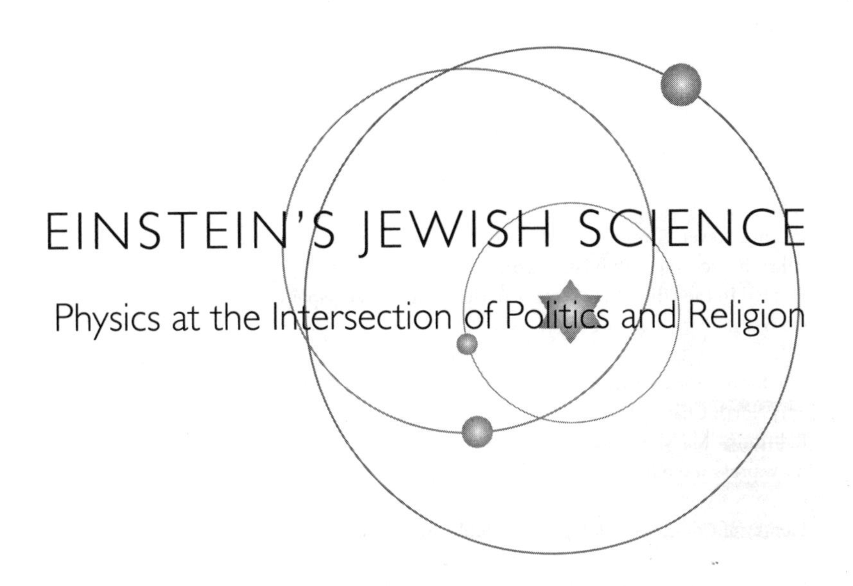

EINSTEIN'S JEWISH SCIENCE

Physics at the Intersection of Politics and Religion

Steven Gimbel

THE JOHNS HOPKINS UNIVERSITY PRESS
Baltimore

Printed in the United States of America on acid-free paper
9 8 7 6 5 4 3 2 1

The Johns Hopkins University Press
2715 North Charles Street
Baltimore, Maryland 21218-4363
www.press.jhu.edu

Library of Congress Cataloging-in-Publication Data

Gimbel, Steven, 1968–
Einstein's Jewish science : physics at the intersection of politics and religion / Steven Gimbel.
p. cm.
Includes bibliographical references and index.
ISBN-13: 978-1-4214-0554-4 (hdbk. : alk. paper)
ISBN-10: 1-4214-0554-7 (hdbk. : alk. paper)
ISBN-13: 978-1-4214-0575-9 (electronic)
ISBN-10: 1-4214-0575-X (electronic)
1. Relativity (Physics)—Philosophy. 2. Einstein, Albert, 1879–1955—Philosophy. 3. Jewish Science. I. Title.
QC173.55.G55 2012
530.092—dc23 2011036254

A catalog record for this book is available from the British Library.

For my grandparents,

Fran and Milt Gimbel

and Betty and Leonard Gallant

CONTENTS

EINSTEIN'S JEWISH SCIENCE

Einstein's Jewish Science

If my theory of relativity is proven successful, Germany will claim me as a German and France will declare that I am a citizen of the world. Should my theory prove untrue, France will say that I am a German and Germany will declare that I am a Jew.

—Albert Einstein to the French Philosophical Society, April 6, 1922

Sometimes even the cynics aren't cynical enough. Despite its success, indeed because of it, Albert Einstein's theory of relativity was denigrated by Nazi sympathizers, who dismissively labeled it as "Jewish science" during the run-up to World War II. To their chagrin, Einstein was succeeding in radically revising the scientific community's understanding of space, time, matter, energy, motion, and gravitation.

Of course, relativity—like all revolutionary ideas—has had more than its share of detractors, a few hanging on until this very day. Many of the supposed problems with the theory are what philosopher Hans Reichenbach, one of Einstein's earliest defenders, called "off-key notes," that is, uninformed objections from opponents who failed to understand the theory.[1] It is not unusual for physicists today to receive letters and e-mails from nonscientists with a new "proof" that Einstein is wrong.

But some objections came from the world of physics in which renowned authorities bristled at the thought of an unknown patent clerk with no Ph.D. undermining concepts such as absolute space, absolute time, and luminiferous aether that had been the standard fare of physical theory for generations. These critics included some of Einstein's personal heroes: the German physicist and father of positivism, Ernst Mach; the French polymath, Henri Poincaré; and the Dutch theoretician, H. A. Lorentz. Their concerns would seem unremarkable in normal scientific discourse: lack of supporting experimental evidence, the existence of competing theories, or the historical fruitfulness of concepts Einstein rejects. Over time, these criticisms were successfully addressed and relativity became so strongly accepted that it is now viewed as part of "classical" physics.

But then there were other adversaries. Philipp Lenard, winner of the Nobel Prize in physics in 1905, became one of the leading spokesmen for the Aryan physics movement that sought to cleanse German science of what it saw as its modern corrupting Semitic influences. In the foreword to *German Physics*, a four-volume encyclopedia of Aryan science, he wrote:

> "German physics?" you will ask.—I could also have said Aryan physics or physics of the Nordic type of peoples, physics of the probers of reality, of truth seekers, the physics of those who have founded scientific research.—"Science is international and will always remain so!" you will want to protest. But this is inevitably based upon a fallacy. In reality, as with everything that man creates, science is determined by race and blood . . . Nations of different racial mixes practice science differently.[2]

Einstein's theory of relativity, the proponents of Aryan physics argued, was not only wrong, but in steering physics away from the true German approach, it was also malignant.

Nowadays, we wave off the claim that relativity is Jewish science as just another example of the intellectual bankruptcy and ideological absurdity of National Socialism. We are quick to dismiss this claim, as we are of all others made by Hitler and his minions, in part because we know them to have been merciless propagandists seeking to advance their dehumanizing agenda through psychological, as well as actual, warfare. We therefore easily lump together all Nazi arguments into the single category of sociopathic nonsense.

But while some of the arguments offered in defense of the theory of a master race were surely made for marketing purposes only, others were proposed seriously and need to be honestly debunked lest they arise again clothed in new robes. We know that great minds like Nobel laureate Lenard, famed mathematician Ludwig Bieberbach, and Martin Heidegger, one of the most influential philosophers of the twentieth century, all stood shoulder to shoulder with their Nazi brethren. Such minds cannot be simply dismissed without peril.

It is understandable why we paint all National Socialist arguments with such a broad brush. There is moral and intellectual comfort to be had in taking all Nazis to be inhuman creatures driven by an irrational ideology based on completely erroneous assumptions, their philosophies devoid of any deep and legitimate intellectual work requiring serious engagement. Summary dismissal signifies that the Nazis, their sympathizers, and enablers were not like us and that we are not like them. Surely, we would not have allowed the crimes that they perpetuated to have happened, or worse, participated in them ourselves. We are thus safe from their influence, insulated in the same way that they wished to be from ideas that make us uncomfortable by banishing them with the same oversimplifying tactics that they practiced.

For our own peace of mind, however, we may be quick, too quick, to write off the claims made in defense of Nazism's ghastly creed. This is not to argue for any sympathy toward Nazism; rather, it is to say that perhaps there are insights to be gleaned from a deeper investigation of certain claims made in the advance of it. Could there be new seeds growing from the intellectual dung heap; flowers that can be transplanted into more nourishing and fertile soil?

Beyond our instinctive rejection of any argument associated with Nazism, another reason we often dismiss as nonsense the claim that Einstein's theory of relativity is "Jewish science" is that such a view violates our basic understanding of the nature of science itself. Science is the search for absolute laws and empirically adequate theories that can be derived and tested by anyone. The world is how it is and our proposed scientific descriptions of it may be true or false, may give us better or worse predictive models, may be intuitively satisfying and conceptually elegant or not. But all of our means of evaluating disparate scientific theories, all of the ways we determine which of a set of competing explanations of natural phenomena is superior, are—and should be—wholly independent of religious doctrine or the ethnic, cultural, or familial background of the theory's author. Science is objective, and this objectivity renders the phrase "Jewish science" oxymoronic.

How strange, then, that one of the most forceful advocates of the theory of relativity would go on to argue that

> Jews are a group of people unto themselves. You can see their Jewishness in their appearance and notice their Jewish heritage in their intellectual work and perceive a profound connection between their nature and the numerous interpretations they give to that which they think and feel in the same way.

Perhaps stranger still is that the author of this argument that Jewish qualities might be inherent and recognizable in the intellectual work of Jews is none other than Albert Einstein himself.[3] Einstein's own words suggest that we must take seriously the possibility that the Nazis were in some sense correct about his theory. Maybe relativity *is* "Jewish science" after all.

Since the pioneering work of Robert K. Merton in the 1940s, sociologists of science have argued that we cannot understand the advancement of science without understanding the ways scientists interact with each other, the hierarchies and institutions they create, and the ways in which they influence and are influenced by larger social, political, economic, and even the religious forces at work during their time.

Merton argues persuasively that the scientific revolution of the seventeenth century, for example, is inseparable from the theological and political wrangling between the weakening Catholic Church and the rising tide of Protestantism.[4] Virtually all of the great thinkers of this period—Isaac Newton in physics, William Harvey in biology, Robert Boyle in chemistry—were Protestant, which is surely no coincidence. The intellectual progress these scientists represent was in large part a revolt against the prescribed infallibility of the Catholic Church. Not only did the new scientific theories differ from the long-entrenched writings of Aristotle that were sanctioned by the Church but the new inductive methodology itself was also consistent with the antiauthoritarian theology of the Protestants. In the same way that one's relationship with the Divine need no longer be mediated through the Church's authoritative dictates, the blueprints of God's Creation were freed from the grip of the Church and open to scientific investigation by anyone. It is impossible to understand fully the advance of empirical science during this period without seeing its relation to the religious commitments of its practitioners.

So, if the great advances of the scientific revolution can be meaningfully called "Protestant science," might not Einstein's revolutionary work be considered "Jewish science" in a similar fashion?

In the Jewish community worldwide, few approach the stature attached to Einstein. There is great delight in pointing out that Dinah Shore, Abe Fortas, William Shatner, Marc Chagall, Felix Frankfurter, not to mention both Simon and Garfunkel, all three Stooges, and all four Marx Brothers are Jewish. But among famous Jews, Einstein is in a category unto himself.

True, he was a committed Zionist as well as a scientist, but compare his standing with that of Chaim Weizmann, who is revered as a Zionist leader even by those who have no sense that he was a famous biochemist whose discoveries played a significant part in the Allies' victory in the Great War.[5] Einstein, in contrast, is a revered Jewish scientist who happened to be a Zionist. It is Einstein's theories, it is his scientific work that make Einstein, Einstein. But if nothing about his work, especially the theory of relativity, is Jewish, then we have no choice but to reduce Einstein to the level of Adam Sandler.

But, of course, he is not just another Jewish celebrity. There is pride that Sandy Koufax, one of the greatest left-handed pitchers in the history of baseball, was Jewish. But there is a difference in the status of Einstein, one of the greatest scientists in history, and it isn't only because science is held to be more important than baseball. It is a more substantive difference. African Americans, for example, are proud of Jackie Robinson in the same way Jews are proud of Einstein.

Now, there was nothing about the way the great number 42 threw, hit, or ran that was particular to African Americans, but Jackie Robinson is Jackie Robinson because he was an *African American* baseball player. No one wants to deny that Jackie Robinson was black while he played and that it was a factor on the field. But we easily grant that Einstein was in some sense Jewish when he walked into the office in the morning and that he was a Jew when he walked home at night, but we are reticent to say that he was a Jew in between when he was sitting at his desk, pencil in hand. How can we reconcile the fact that when it comes to his work we make him check his Judaism at the door and yet at the same time think of him as the Jewish Jackie Robinson? Why can't we hold that in the same sort of sense in which Jackie Robinson was an African American ballplayer, Einstein is a Jewish scientist? And if we can hold this, what precisely would it mean about the way we understand his science or the way that it was done that justifies this privileged place?

The pride contemporary Jews feel in claiming Einstein as a member of the tribe, however, is dwarfed by the hatred of his nationalistic critics whose attacks on the theory of relativity need to be understood in historical context. Germans looked on enviously over a period of centuries as virtually every other nation in the region—the British, the French, the Dutch, the Spanish, the Portuguese, the Austrians, the Turks—had enjoyed the role of dominant world power. But by the mid-nineteenth century, Germany had not only been unified, it had matured, and the cultural, intellectual, and scientific advances it produced yielded the sense that the time of German ascendance had finally arrived. The works of Ludwig von Beethoven and Richard Wagner were masterpieces, conveying strength and assuredness in their unabashedly Romantic fashion. Friedrich Hegel proclaimed German culture the pinnacle of human evolution while Friedrich Nietzsche philosophized with a hammer. The emergence of individuals like the prominent physicist Hermann von Helmholtz and the rise of a new technological economy provided evidence that history was building toward a crescendo whose climax was German intellectual and industrial dominance.

This sense of cultural superiority was all but shattered by Germany's devastating military defeat at the end of the Great War. Conservatives were loath to even admit that the war had been lost. After all, to lose a war is to be conquered and no one had even threatened to invade the Fatherland, much less occupy or seriously challenge its sovereignty from without. But if the war had been lost, it was all the fault of the liberals, socialists, and communists who had been insufficiently patriotic. In the wake of this humiliation, a deep political schism became manifest as the right, on the one hand, sought to maintain the structure of old, believing that history owed Germany its role as the great European empire, a mantle so close, a title so deserved, a position it had earned. The political left, on the other hand, wanted to end the age of empires altogether. No one should be a dominant world force. This instinctual internationalism of the left was a slap in the face to all that the nationalists held true.

Einstein was the public face of this movement. His hatred of German militarism led him to renounce his citizenship at age sixteen in 1895, after his father's electronics business went bankrupt and his family moved to Italy. He had been left behind to finish his schooling, but he detested the German mode of instruction at the Luitpold Gymnasium, calling his teachers "lieutenants" and the schoolhouse "barracks." Convincing the family doctor to write him a note excusing him from school for a nervous breakdown, he left Germany for good . . . or so he thought.

Years later, Max Planck convinced him to return to a prestigious position at the University of Berlin, where he would be surrounded with colleagues he liked and admired. He would go back, yes, but he would ignore his own advice to keep his mouth shut. As world war loomed, he again saw everything he had abhorred about German militaristic nationalism and would let his feelings be known at a time of patriotism when his sentiments were seen as less than friendly. His lifelong contempt for all that conservatives held to be great and right made them natural adversaries. Our cartoon image of the white haired, grandfatherly, wise clown is the opposite of young Einstein who was overtly political, edgy, sharp-tongued, and unafraid of controversy. In the war that was to establish Germany's rightful place, he had been a pacifist speaking out for international unity and cooperation when conquest was the order of the day.

In the unsettled peace that followed, it was this traitor to the Fatherland, as conservatives saw him, who was held up as the example of the good German, the rehabilitated German. He willingly visited the enemy and gave speeches with unabashed thumbs in the eye of the German nationalists he so detested. In 1921, he sailed to the United States with Chaim Weizmann to raise money for the Hebrew University in Jerusalem and for himself (his divorce and the German inflation caused him financial problems). He gave high-profile speeches at Columbia and Princeton and then moved on to England where he spoke at King's College, London, and Manchester University. A few months later in 1922, he spoke in Paris at the Sorbonne. He sailed aboard ships under the Allied flags, received warm embraces from the enemy, and attacked German nationalists in his public addresses and highly visible interviews with them:

> Eastern European Jews are made the scapegoats for certain defects in present-day German economic life, things that in reality are painful aftereffects of the war. The confrontational attitude toward these unfortunate refugees, who have escaped the hell that Eastern Europe is today, has become an efficient and politically successful weapon used by demagogues. When the government contemplated measures against Eastern European Jews, I stood up for them in the *Berliner Tageblatt*, where I pointed out the inhumanity and irrationality of these measures.[6]

To brazenly air dirty laundry as Germans suffered was enraging, but to do so in the presence of those who had reduced the proud German nation to the role of debtor and had killed its loyal sons in battle was a betrayal of the vilest sort, and loyal patriots seethed.

As a result of reparations and a weak central government, the suffering of ordinary Germans was pervasive. The nationalists saw this coupled with, or indeed a result of, an attack on German culture. Where Aryan Germans had been the leading lights in the worlds of mathematics and science, art and industry, suddenly it had all changed and Jews were now claiming the glory. The Weimar government, which the Aryans saw as illegitimate, placed Walther Rathenau, an ostentatious Jew, as foreign minister, as it dismantled the anti-Semitism built into German law. The Romantic lyricism of Franz Schubert and Johannes Brahms was being eclipsed by the overly intellectual Gustav Mahler and Arnold Schoenberg. Philosophy went from Hegel's claims of German culture as the end of history to abstract logical symbolism or individual experience because of thinkers like Hans Reichenbach and Edmund Husserl who had Jewish backgrounds. Hugo Dingler, a Nazi academic, complained to authorities that "because of their cleverness, speed, and sure memory power, Jewish mathematicians are everywhere in the foreground,"[7] having taken prominent places at the University of Göttingen, the mecca of mathematics. In every field, Jews appeared to gain what Aryans lost. It was a small step to resentment and then only short slide into talk of conspiracy. And there stood Einstein, the symbol of it all, never missing a chance to provoke.

It was on this ground that the Aryan physics movement was born. The greatness of German culture would be its flowering as the ultimate expression of classical thought, which needed only to be polished anew, not cast aside completely. The new and bizarre scientific methodology replaced observation and intuition with artificial mathematical contrivances. Science was seduced away from "being in the world," a concept stressed in the writings of Friedrich Nietzsche and Johann Wolfgang von Goethe. The Jews replaced reality with formalism, and science became nothing but a game of symbolic manipulation. Relativity, they argued, was not a mistaken physical theory, but a philosophical problem, a methodological fallacy. The less than noble basis of these novel, strange, and counterintuitive ideas that were being advanced by this new generation of physicists needed to be exposed. In doing so, the similarity between this mathematical approach to science and the rabbinic method of interpreting sacred text with its multiple perspectival meanings was not lost on the advocates of Aryan physics.

The work of Einstein, the political firebrand, the cultural heretic, the arrogant traitor, would be attacked in a way that would restore the rightful place of true, proper German science and would allow once more for the ascendance of the German spirit, *Geist*. By repudiating his theory, Einstein's power would be drained. As went his science, so would go Einstein. The theory of relativity, the proponents of Aryan physics argued, is not true science but "Jewish science."

BUT IS IT? WE CAN ONLY RESPOND to this charge if we first answer the question "What is 'Jewish science'?" In the same way that you can only tell whether an electron is a boson if you first know what a boson is,[8] we cannot simply wave our hands and dismiss as absurd and ineffective Nazi propaganda a catchphrase designed to smear Einstein's work until we work out clearly and explicitly what this phrase means. In fact the question, "What is 'Jewish science'?" turns out to be a very Jewish question.

To investigate its several possible meanings leads us to further discussions and questions. In the first place, "Jewish science" relates to the identity of the researcher, that is, a theory is an instance of Jewish science if the scientist who developed it is Jewish. Clearly, this is a weak notion that in no way requires the theory itself to be related to or have its origin or influences in Judaic ideas or beliefs. Based on this interpretation, the theory of relativity is in fact Jewish science if Einstein himself was Jewish. Of course, to determine whether Einstein was a Jew, we need to know what makes one a Jew and whether Einstein would satisfy the condition. Einstein stated explicitly that he did not believe in the God of Abraham, yet he also unambiguously referred to himself as a Jew. Even at this very basic level, Jewish identity, simple as it may sound, is a question of considerable and longstanding debate.

The second interpretation we might explore when we hear the words "Jewish science" is whether a theory is Jewish science if it is influenced in significant ways by, makes inextricable reference to, or is derived from Hebraic beliefs or traditions. This will require examining the historical context and influences at work upon Einstein when he developed the theory of relativity to determine if his work can be traced specifically to biblical or Talmudic sources.

In addition to looking at the content of the theory itself, we can also examine the style of reasoning that brought Einstein to it. If we examine the differences between Catholic, Protestant, and Jewish approaches to thinking about the sacred, can we find structural similarities to the differences between the theories of René Descartes and Isaac Newton about mechanics and gravitation, on the one hand, and those of Einstein, on the other? Can we see in the scientific methodology of these physicists analogues to the mode of theological deliberation in their respective traditions?

Beyond these considerations, if we consider what the phrase "Jewish science" meant to a German speaker between the world wars, it turns out that the term "Jewish" was loaded with all sorts of cultural baggage. With the abdication after the war of Kaiser Wilhelm, beloved leader of the conservative end of the German political spectrum, and the subsequent rise of the Weimar Republic, the term "Jewish" took on a very clear political meaning. It conveyed liberal sympathizer, a role Einstein relished, but it likewise implied notions contrary to traditional German values. It was not only Aryan physics advocates like Lenard who asserted that science is pregnant with politics, contemporary sociologists made similar arguments.[9] So, are there political suppositions underlying the theory of relativity?

While this influence can be viewed from the lens of political principles it also can be seen from the point of view of intellectual history writ large. Einstein's theory appeared in 1905, during a period of upheaval across the intellectual spectrum from mathematics, science, and the social sciences to the humanities and the arts, the same year that Freud published his major works on psychoanalysis and sexuality, Picasso entered his rose period, and Mahler scored his seventh symphony. In virtually all of these simultaneous revolutions, prominent Jews led the way, and thus the term "Jewish" implied the rejection of classical forms and norms across the board. Engineer and founder of the Association of German Scientists for the Preservation of Pure Science, Paul Weyland, called the theory "scientific Dadaism," explicitly linking Einstein's work to modern art. Is there, in fact, a coherent unifying intellectual theme to these currents and does relativity fit neatly into it? Was relativity "Jewish science" in this way?

Finally, we might ask, was Einstein's theory perceived to be "Jewish" among the contemporary Jewish community with scholars and debates of its own? Figures like Martin Buber, Erich Fromm, and Walter Benjamin are admired across disciplines, but in fact, many of the insights of their widely acclaimed works come out of conversations that centered around explicitly Jewish themes and questions. As they and their colleagues were often engaging Judaic concerns, did Einstein's work, consciously or not, have an impact on what Jewish scholars discussed or how they discussed it?

So, to return to the place from which we started: Is the theory of relativity "Jewish science"? Recall the old joke, "Why do Jews always answer a question with another question?" The answer, of course, is "Why shouldn't we?" And to determine whether the theory of relativity is indeed Jewish science, we must, necessarily, go on to ask many more questions.

EINSTEIN LOVED TO TAKE LONG WALKS in the Alps with friends while discussing physics. Like those meandering trails, the discussions here will take wide loops and many turns into areas we might not have expected to visit on our way. Your indulgence is begged on these winding intellectual paths as we try to determine from many different directions whether there is any meaningful sense in which the theory of relativity could be thought of as "Jewish science."

CHAPTER ONE

Is Einstein a Jew?

Perhaps the easiest way to undermine the Nazi claim that Einstein's theory of relativity is "Jewish science" would be simply to say that Einstein wasn't Jewish. Surely, it's absurd to consider the work of a non-Jew to be Jewish science. End of story.

So, was Einstein Jewish? The Nazis thought so. Jews think so. Einstein himself even thought so. Of course, what each mean by "Jew" is something completely different. What seems like a simple, straightforward task—determining what makes someone a Jew—is actually quite tricky.

Most Jews would answer this question by appealing to the traditional tribal definition according to which being Jewish simply means being born of a Jewish mother. It is an ancient requirement, which makes it all the more puzzling. Judaism, especially early on, was hardly a form of protofeminism. The patriarchal roots are clear in contemporary Orthodox synagogues where women are segregated from men, cannot become rabbis, and must cover their hair in public. In Israel, women may not pray like men at the Western Wall, Judaism's holiest site. Why make the woman, then, the key to membership in the tribe?

This is made stranger still when we realize that the privileging of the masculine contribution to the Jewish community is also applied to the masculine contribution to the procreative process. In Genesis,

Onan is killed for displeasing God when he spilled his seed upon the ground instead of impregnating his brother's widow. This story is often taken as an allegory whose meaning is a prohibition on masturbation—male masturbation, that is.

Women, in contrast, are not constrained in the same way. In Leviticus, menstruating women are commanded to be separated from others and are banned from marital relations, but the passing of the monthly period of fertility without becoming pregnant is not problematic in the same way that wasting semen is. Of course, the notion of an ovum, the woman's egg, was unknown to the ancients, who held semen to create the white parts (e.g., bones, teeth), whereas the woman's menstrual blood produced the red (e.g., tissue, organs).[1] With this theory of propagation, why then place the identity of the offspring within the body of the woman?

For much of the history of Judaism, technology was not sufficient to give those with wealth and political power control over the masses. Think of the Pharaoh's fear in the first chapter of Exodus that the Jews had reached a sufficient size that they posed a threat to his power if they chose to ally with his enemies in a time of war. Then, more than now, size mattered. The arms race at that time was literally a race to produce more human arms. "Go forth and multiply" wasn't merely a command for family values to the ancient Hebrews; it was a prime directive to allay the perennial fear that the tribe would be wiped from the face of the Earth. Masturbation and homosexuality? Wasted opportunities to grow the ranks.

Why then place the clan membership card in the womb of the mother? The last of the ten plagues, the killing of the firstborn males, was not an idea pulled from thin air. It was a horrible, perhaps the ultimate, harm that could be done to a people. Killing the boys of a rival eliminated the defeated tribe's chance of regaining martial force. But in addition to murdering young men, to the victors went the spoils. In raping the women, some would inevitably become pregnant. By considering the mother to be the source of belonging, these offspring would be Jewish, a protection from eradication. It is purely a pragmatic definition linked to the survival of the community.

Some scholars argue that this line of reasoning dates from the period of the early Diaspora, as early as the eighth century BCE, when Jews were nomadic and the size of tribe was a constant concern. Others place it later in Roman times when the rape of Jewish women was not unusual given the collection of the "Jewish tax" levied on non-Romans. Advocates of this view cite the similarity of Roman law and rabbinic law of the period, both holding that if someone from within the society marries a stranger from outside, it is the mother's status that the offspring inherits. Still others claim that Jews acquired it during the earliest of recorded times, since we know that ancient Egyptian and Mesopotamian societies also had matrilineal definitions of identity. Whatever its origin, being "Jewish" because one has a Jewish mother has become the entrenched notion within the community.[2]

So, by this ancient criterion, was Einstein Jewish? Yes. Einstein's mother, Pauline Koch, had deep roots in northern German Jewish society. She experienced a standard Jewish upbringing and, while the house she kept was secular, the Einstein family lived very much in the context of German Jewish culture. Indeed, Einstein's mother was more than Jewish; she was in many ways the stereotypical *yiddishe Mame*—strong-willed, outspoken, and more than willing to employ guilt to get what she wanted from her children, especially with her son, Albert. When Einstein, for example, refused to break off a relationship with a woman his mother disapproved of—Mileva Marić, the woman who would be his first wife—she used all the tools she could muster to change his mind. In a letter to Marić, he describes the scene:

> Mama threw herself onto the bed, buried her head in the pillow, and wept like a child. After regaining her composure, she immediately shifted to a desperate attack: "You are ruining your future and destroying your opportunities." "No decent family will have her." "If she gets pregnant, you'll really be in a mess." With this last outburst, which was preceded by many others, I finally lost my patience.[3]

IF EINSTEIN WOULD BE CONSIDERED Jewish by the definition used within the community, what about definitions from outside? The

Nazis certainly would not give Jews the power of self-definition, but it was a task they struggled with. Arriving at a strict definition of "Jew" was important, since it is imperative, after all, to know exactly who you are persecuting before beginning a systematic reign of terror against them.

So, within weeks of Hitler's rise to power in 1933, the *Berufsbeamtengesetz*, the Law of Restoration for Career Civil Service, went into effect, barring Jews and enemies of the regime from prestigious, well-paying government jobs. The legislation said that "civil servants who are not of Aryan descent are to be retired; if they are honorary officials, they are to be dismissed from their official status."[4]

While this law sought to purge Jews from important positions, there is no explicit definition of the term "Jew" in this document, so they had to produce a second one, a regulation for implementation of the law. Here, we find exactly what we would expect from those who sought racial purity:

> A person is to be considered non-Aryan if he is descended from non-Aryan, especially Jewish, parents or grandparents. One parent or grandparent is sufficient to make one non-Aryan. This is particularly true in cases in which a parent or grandparent was of the Jewish faith.[5]

For the Nazis in 1933, any Jewish blood makes you Jewish.

But this criterion ran into some difficulties. One of the defining events for German nationalists was the Great War. It was to be the time of the ascendance of German greatness upon the world stage. Those who fought, especially those who died, were revered as great patriots.

When Hitler's rise made some in Germany nervous, Paul von Hindenberg was made president to try to act as a buffer, to soften Hitler's power. Von Hindenberg was seen as the only person who could do this, because his stature and nationalistic credentials were beyond question as a former trusted chief advisor to the Kaiser and as the field marshal who led the great victory over the Russians at Tannenberg.

Even Hitler knew that von Hindenburg's social cachet had to be respected. So, when von Hindenberg insisted that the law as written was unacceptable, it was altered in accord with his wishes to make exceptions for some Jews:

> If a civil servant did not already have civil-service status on August 1, 1914, he must prove that he is of Aryan descent, or that he fought at the front, or that he is the son or father of a man killed in action during the World War. Proof must be given by submission of documents of this includes documentation (birth certificates, marriage licenses of parents, military documents).[6]

These restrictions remained an annoyance to the Nazis and were eliminated the next year following von Hindenburg's death.

A year later, the *Reichsgesetzblatt*, the Reich's Citizenship Law, passed in September of 1935, which was designed to not only remove Jews and other undesirables and political opponents from governmental posts but also to strip them all of citizenship, including the right to vote. It, too, needed instructions for implementation and in this document of November 1935 they take a second pass at defining "Jew":

> A Jew is anyone descended from at least three grandparents who are fully Jewish as regards race.[7]

where

> A grandparent is deemed fully Jewish without further ado, if he has belonged to the Jewish religious community.[8]

There are two important elements in this definition. First is the uneasy relation here between Judaism as a race conditioned by blood and as a religious community bound by common history, customs, and rituals, and thereby not a necessary result of biological elements. The Nazis' embrace of eugenics was not metaphorical—something that both delighted and concerned German scientists who lead a strong eugenics project before 1933.[9] In positing a true connection between race and blood, the Nazis sponsored a vigorous program in serology, the study of blood.[10] So, the move to defining a biological notion of race in terms of a cultural factor became a real problem for Nazi race theorists.

The second is that there is a weakening of the definition of "Jew." Where inclusion in 1933 was a matter of having any Jewish heritage at all in your background, in 1935 we see a distinction between the general notion of Jew, the concept of racially full Jew, and a new group, *Mischlinge*, or mixed Jews, who have fewer than three Jewish grandparents.

> §5. (2) Also deemed a Jew is a Jewish *Mischling* subject who is descended from two fully Jewish grandparents and
>
> a. Who belonged to the religious community when the law was issued or has subsequently been admitted to it;
> b. Who was married to a Jew when the law was issued or has subsequently married one;
> c. Who is the offspring of a marriage concluded by a Jew . . .
> d. Who is the offspring of extramarital intercourse with a Jew . . . and will have been born out of wedlock after July 31, 1936.
>
> §6. (1) Requirements regarding purity of blood exceeding those in §5 that are set in Reich laws or in directives of the National Socialist German Workers Party and its units remain unaffected.[11]

Mischlinge of the first degree, those with only one Jewish grandparent, are now considered Aryans while those with two grandparents, *Mischlinge* of the second degree, qualify as Aryans as long as they don't satisfy any of these other conditions.

This is puzzling. Having cemented power and crafting a policy that removes crucial rights from all Jews, not just those with government jobs, why would the Nazis now move to a looser definition of Jew?

It turns out that there are several reasons for this. First, this is in part a political compromise. The German government at this point was a combination of new appointments by the Nazis who held positions within the party, owed their jobs to the party, and were radical believers in Hitler's vision, on the one hand, while on the other were lifelong German bureaucrats whose views were not necessarily in line with National Socialism but whose years of commitment, expertise, and "pure blood" made them appropriate for the positions they held. While the law was a victory for the Nazi faction, it also was influenced by the more moderate elements in the Ministry of the Interior.[12]

Second, there were *Mischlinge* that the Nazis wanted to lay claim to, Germans who had done significant things and others who could go on to serve honorably in the military (recruits were desperately needed) who happened to have some small amount of Jewish ancestry.[13] They could justify this by claiming that it was the non-Jewish part of them that was responsible for the qualities that allowed them to ascend to greatness.

Hitler himself worried that he might be a *Mischling* of the first degree due to his paternal grandfather whom Hitler never knew. What he did know was that his father was born out of wedlock, but the father of his father was not explicitly documented. Writing in Nuremburg before his execution, Hitler's legal advisor and friend Hans Frank claimed that Hitler tasked him with finding out the identity of his grandfather and that his research showed that he was likely Jewish. Frank thought he had found that Hitler's grandmother had been a domestic servant for a Jewish family in Graz and that their nineteen-year-old son had impregnated her.[14] Historians have since shown that Frank's research is riddled with errors and his conclusion is highly unlikely. But whether Hitler was, in fact, a *Mischling* is not the point, merely that it was something he seriously considered that would have had to have had an effect on his thinking about German racial identity.

The *Mischling* problem was not merely a pragmatic issue. It was one that confounded Nazi race theorists. They had come face to face with the problem of the chessboard—is it a white board with black squares or a black board with white squares? Was someone with two Jewish and two Aryan grandparents a member of the Aryan race whose blood was polluted by Jewish contaminants or was the person a Jew who would be elevated above other Jews because of the power of the master race?[15]

These questions seem odd to us, but when absolute racial distinctions provide the basic categories from which your entire worldview emerges, these questions become meaningful. Thomas Kuhn argued that language is a communal phenomenon and that we all work within "paradigms," sets of rules and invisible presumptions that are imminent within a linguistic community.[16] Our language shapes and constrains what and how we think because our language provides us with the atomic concepts we use to make sense of the world, concepts that strangers to our linguistic community do not possess. The uninitiated are often baffled at the move of the knight on the chessboard, but to those who are able to think in the chess language, nothing could be more natural. "Of course, that's how the knight moves." "Why?" "Because that's just how it moves."

In the same way, absolute racial distinctions made perfect sense to the Nazis. Further, it necessarily followed from their basic conception that these races could be ranked according to degree of perfection. But this gave rise to a philosophical conundrum, a puzzle in the paradigm. If race is in the blood, then one drop of Jewish blood should be enough to make one Jewish. But then that would make Jewish blood more powerful than Aryan blood. Yet, Aryan blood is superior and therefore stronger. So, how could Aryan racial superiority be made consistent with the desire for purity of blood? The appearance of the category *Mischling* is, in part, an attempt to finesse this problem.

But this new definition did no good for Einstein. His mother, Pauline Koch, was the daughter of Jewish parents Julius and Jette Koch, while his father, Hermann Einstein, was the son of Abraham and Helene Einstein, who were both members of the longstanding Jewish community that had been settled since the middle of the nineteenth century in the town of Buchau, near Ulm, where Albert was born. By the criteria of his time, since all four of Einstein's grandparents were thus Jewish, so then was Einstein.[17]

BUT MANY OF THOSE CLASSIFIED AS JEWISH by the Nazis were shocked by it because they did not consider themselves to be Jews. They had converted, many even changing their names to something that did not sound Jewish. Jews in Germany and Austria at the time had a much stronger sense of felt cultural belonging through their citizenship than through their religious heritage. They thought of themselves primarily as Germans and Austrians, not as Jews.

This is epitomized by Lise Meitner, the first person to realize where Einstein's famous equation, $E = mc^2$, was applicable in nature and who played piano to Einstein's violin at Nobel laureate Max Planck's home, when Planck, himself an accomplished pianist, declined to play.[18] Born to a secular Jewish family in Vienna, Meitner quickly showed herself to be brilliant, sleeping with a mathematics text beneath her pillow at age eight. In 1905, for her dissertation showing that the way some materials conduct electricity is related to how they conduct heat, she became just the second woman to receive a Ph.D. from the University of Vienna. But the doctorate was hardly a guarantee of an academic post, something no woman had ever achieved in Austria.

She tried teaching at a girls' school, but it did not suit her. She worked informally for Stefan Meyer, a physicist whose interest was in radioactivity, the same field as the most successful female physicist of the time, Marie Curie, who had the advantage of a well-established husband with which to face the discriminatory world of turn-of-the-century science. In Meyer's lab, Meitner did good work on the scattering of the mysterious alpha particle, publishing her findings in the prestigious *Physikalische Zeitschrift*.

Meitner then did two things that helped her attain an academic position. First, she left Austria for Germany where, on September 28, 1907, she met Otto Hahn, who was also working on radioactivity and who quickly invited her to work as his assistant while she attended Planck's lectures. Although she was hired to work in Hahn's lab, she couldn't actually go there because women were not allowed anywhere in the chemistry building out of fear that their hair would catch fire. Hahn negotiated a compromise in which Meitner could enter a carpenter's shop in the basement of the building where he had already started his experiments. The second thing she did was to convert. A year and a day after she met Hahn, she was baptized into the Evangelical Protestant church.

Meitner and Hahn, who would come to refer to each other as brother and sister, worked together as equals and published prodigiously, studying beta radiation, which was even less well understood than alpha. When their work got Hahn an invitation to join the well-funded Kaiser Wilhelm Institute for Chemistry, Meitner was also invited but only as an unpaid guest. Despite her growing reputation and cutting-edge contributions, she had no job until Planck appointed her as an assistant professor under him, the first woman in Prussia to attain such a post. The stature of Planck and her association with him finally led to her appointment as a legitimate associate at the Kaiser Wilhelm Institute alongside Hahn in 1913.

The fact that the appointment was made by Planck spoke volumes, as he was not only a scientific mind of historic proportions but he was also well-known for being conservative in both politics and manner. While he and Einstein shared deep mutual affection and admiration, the two were as different as could be imagined. Einstein despised the rigidity of the German schooling, while Planck thrived under the unyielding discipline. His clothes were starched as stiff as he was, working for hours standing at a high desk, his daily constitutional and piano practice scheduled to the minute. Planck was a dedicated nationalist whose interest was in furthering science, not women's rights. So, when Meitner presented herself, he certainly was not seeking to make political waves, simply to hire a talented assistant.

With the outbreak of the Great War, Meitner asked the government to allow her to turn the Kaiser Wilhelm Institute into a military hospital, but when that failed she used her expertise to become a nurse and x-ray technician. As a member of the Austrian army, she served at a facility near the Russian front where she confronted the horrors of the wounded and dying, as Hahn was recruited by Einstein's dear friend Fritz Haber to help with his efforts developing the first chemical weapons.

After the war, Meitner went back to work with Hahn and then in her own laboratory at the Kaiser Wilhelm Institute. Both continued to climb the academic ladder and, when Hahn left his place as head of the Institute to spend a year at Cornell in 1933, Meitner was put in charge temporarily his stead. This prominent position, her commitment to Germany and the war effort, not even her conversion to Protestantism could save her from the encircling Nazis. The combined political pull of Hahn and Planck were not enough to stave off the purge, and the Nazis revoked her credentials at the University of Berlin under the authority of the *Berufsbeamtengesetz*.

She was dismissed from her university position, but since the institute was independent, she could maintain her appointment there. Meitner's Austrian passport was no longer valid and the Nazis refused to give her permission to travel abroad.

In 1938, she escaped to Holland and then Sweden where Hahn sent her the results of his experiments with uranium. From afar, looking only at his data, Meitner realized what Hahn had been able to do. He had documented the splitting of the atom for the first time. The energy he was measuring was derived from mass according to Einstein's equation $E = mc^2$. For the work Hahn, but not Meitner, was given the Nobel Prize in chemistry in 1944.

Meitner's was a legitimate conversion, not a political charade for personal gain. She thought herself a Christian, and most Christians would consider her one as well since being a Christian is a matter of faith, which she professed. But to the Nazis, she was Jewish.

It was something Einstein recognized all too well, viewing these conversions and all other attempts to mollify or impress the German Aryans with great skepticism. German Jews, he contended, had a weakness, an insecurity that led them to try to out-German the Germans, and he loathed it. In 1920, Einstein was invited to participate in a round table about anti-Semitism in the academy from an organization that called itself "German Citizens of Jewish Faith." The name annoyed Einstein in that it made Judaism a faith and implicitly denigrated poorer Eastern European Jews by distinguishing themselves first and foremost as German citizens.

Einstein had left Germany as a teenager with a deep dislike of the nationalistic fervor that was sweeping the country, ultimately landing in Switzerland where he first truly felt at home. The Zurich of Einstein's formative years was an incredibly cosmopolitan place.[19] Jews and women from Russia flocked there for educational opportunities that were denied them at home. His closest friends at the university there hailed from all over central Europe, Romania, Hungary, and Serbia. A prominent professor of Einstein's had emerged from the shtetls of Lithuania. The revolutionary spirit Einstein absorbed as a young man was infused with a multicultural worldview, with the energy and life that comes from a community of strangers seeking a new way.

So when Einstein was invited to address the German Citizens of Jewish Faith, the attempt to elevate itself by stressing its German nature brought with it a dose of Einstein's scorn:

> More dignity and independence in our own ranks! Not until we dare to see ourselves as a nation, not until we respect ourselves, can we gain the respect of others, it must start with us then it will follow . . . Can the "Aryans" have respect for such sycophants?[20]

Attempts to deny one's identity in the face of it will not allow one to avoid oppression, Einstein argued, only leave one oppressed without dignity. To the Nazis, Judaism was a matter of blood not religion.

BUT, OF COURSE, JUDAISM IS A RELIGION. So we could define the property of being Jewish in two other ways, a "theistic Jew" is one who believes in the God of Abraham and an "observant Jew" is someone who carries out the commanded Jewish rituals regardless of belief.

Einstein was raised in an entirely secular home. So much so that his parents sent him to a Catholic school, Peterschule, in Munich where the family had moved the year after Einstein was born, and he was the only boy of Jewish heritage in his class. It was at the hands of his teachers there that he had his first experiences of anti-Semitism. As one who always relished the role of outsider, Einstein determined that if he was going to be seen as the Jewish kid, he would *be* the Jewish kid. He wrote of this period in his autobiography that "I came—despite the fact that I was the son of entirely irreligious (Jewish) parents—to a deep religiosity."[21]

This impulse received an extra jolt when Einstein's parents, concerned about the religious instruction required by the state, brought in an older relative to give him informal lessons about Judaism in order to provide young Albert some sense of his own traditions. For several years, they stuck. He refused pork, "observed religious prescriptions in every detail,"[22] and made up little psalms to sing to himself on the way to school. Young Albert Einstein had become both a theistic and observant Jew.

But a few years later, a new visitor brought Einstein's religious phase to an abrupt end. As a part of the "it takes a shtetl" mentality of the time, it was a custom to have a local Jewish student over for dinner once a week as students, then as now, were often needy, both in terms of money and a good home-cooked meal. Max Talmey was a medical student and regular weekly guest at the Einstein's table. To curry favor with his hosts, he would bring books for their son. One was a geometry text.

> At the age of 12 I experienced a second wonder of a totally different nature: a little book dealing with Euclidean plane geometry, which came into my hands at the beginning of the school year. Here were assertions, as for example the intersection of the three altitudes of a triangle in one point, which—though by no means evident—could nevertheless be proved with such certainty that any doubt appeared to be out of the question. This lucidity and certainty made an indescribable impression upon me.[23]

And so mathematics and science, through the popular books Max Talmey would bring him, became the core of his approach to the universe, turning his beliefs away from the God of Abraham. Einstein later wrote,

> My position concerning God is that of an agnostic. I am convinced that a vivid consciousness of the primary importance of moral principles for the betterment and ennoblement of life does not need the idea of a lawgiver, especially a lawgiver who works on the basis of reward and punishment.[24]

No God, no afterlife. Einstein, at the time he worked on the theory of relativity, was no longer a theistic or an observant Jew.

Yet, Einstein did consider himself religious as an adult. We must be clear that in his writings, the word "religion" is ambiguous. Einstein uses it in three distinct ways. First, religion for Einstein refers to organized religion and its theology. In this sense, we see that Einstein has explicitly and completely rejected the notion of a personal God and was not connected with any religious congregation.

Second, Einstein uses the term as a synonym for morality. Einstein accepted what philosophers call the "fact/value distinction," where there are two completely different flavors of sentences. One type is descriptive and tells us how things are. The sky is blue. Grass is green. Light curves around massive objects. These "fact propositions" comprise the sentences of science. The other kind of sentence tells how things should be, how we ought to act, whether things actually are that way or not, whether we actually act that way or not. Honor thy mother and father. Thou shalt not steal. Don't systematically exterminate minorities and political opponents. These "value propositions" are prescriptive, they tell us what to do, not what we have or have not done.

Following Einstein's beloved David Hume, he thought that we cannot derive "ought" from "is," that is, there is no way to pull out ethical principles from the results of science.[25] No amount of knowledge about the way things are will direct us to the way they ought to be. Since this normative realm is removed from science and since religious systems tend to join behavioral codes to their histories and supernatural elements, Einstein placed morality within the purview of a liberal sense of religion.

This is not to say that he thought that morality was based on the desires of a lawgiver God or indeed on any other metaphysical foundation. Matters of right and wrong were to be understood as "a purely human matter, albeit the most important in the human sphere."[26]

> A man's ethical behavior should be based effectually on sympathy, education, and social ties and needs; no religious basis is necessary. Man would indeed be in a poor way if he had to be restrained by fear of punishment and hope of reward after death.[27]

So, while Einstein's religion is ethical, don't think it pregnant with the sense of religion, the standard notion that Einstein rejects.

Einstein's third sense of the word is his own "cosmic religion," which he sees as the pursuit of the rules by which the universe governs itself, an approach he sees as strongly related to that of seventeenth-century philosopher Baruch Spinoza,[28] a Jew on whom the rabbis of Amsterdam pronounced a *cherem,* the Jewish equivalent of excommunication, for his allegedly heretical views.[29] Spinoza, a lens grinder by profession, saw himself as completing the work of René Descartes who sought to reformulate all human knowledge according to what he called the "geometer's method," using the sort of deductive proofs we find in Euclidean geometry. Spinoza's masterwork, *Ethics,* intentionally looks like Euclid's geometry with numbered propositions, proofs, and explanations about how they are to be derived from definitions, axioms, and postulates.

Logic, for Spinoza, is not merely a matter of style of presentation; rather it is the set of ruling principles that necessarily govern reality. What happens must happen. There is no chance. There is no luck. There is no free will. There is only necessity. This lack of freedom encompasses God as well, who can only do as the logic of his nature commands. God thereby becomes a deterministic slave of logic. This was a view that did not sit well with clergy in either the Jewish or Christian communities.

A pantheist, Spinoza contends that God is not a separate entity who created the universe, rather God *is* the universe. All things are modes of God. The heavens and the Earth, animals, humans, none of it was created by God, rather, all are aspects of God.

> Except for God, there neither is, nor can be conceived (Prop. xiv), any substance, any thing that is in itself and is conceived through itself (Def. iii). But modes can neither be nor be conceived without substance (Def. v). So they can be in the divine nature alone. But except for substances and modes there is nothing (Ax. i). Therefore, everything is in God and nothing can be or be conceived without God, q.e.d.
>
> . . . I have demonstrated clearly enough—in my judgment (corr. Prop. vi and note 2, prop viii), at least—that no substance can be produced or

> created by any other. Next, we have shown (prop. xiv) that except for God, no substance could either be or be conceived, and hence we have concluded that extended substance is one of God's infinite attributes.[30]

We are not individuals created by God with eternal souls to be judged by God or resurrected upon the Earth. We are simply aspects of God's being. It is a fiction to see material objects as discrete, independent things. The world of entities and individuals we commonly speak of—and upon which the Judeo-Christian-Islamic worldview is based—is false. There is only God, God is all, and God is unfolding as God must because of the absolute logic to which he is bound.

As a fellow refugee from Jewish theology with a similar lifelong love of Euclidean geometry, Einstein found much to admire in Spinoza's pantheism. In a telegram to Rabbi Herbert Goldstein that was quoted in *The New York Times,* Einstein said, "I believe in Spinoza's God who reveals himself in the orderly harmony of what exists, not in a God who concerns himself with fates and actions of human beings."[31] Einstein writes eloquently of this belief in his essay "The World as I See It":

> The most beautiful experience we can have is the mysterious. It is the fundamental emotion which stands at the cradle of true art and true science. Whoever does not know it and can no longer wonder, no longer marvel is as good as dead, and his eyes are dimmed. It was the experience of mystery—even if mixed with fear—that engendered religion. A knowledge of the existence of something we cannot penetrate, our perceptions of the profoundest reason and the most radiant beauty, which only in their most primitive forms are accessible to our minds—it is this knowledge and this emotion that constitute true religiosity; in this sense, and this alone, I am a deeply religious man. I cannot conceive of a God who rewards and punishes his creatures, or has a will of the kind that we experience in ourselves.[32]

For Einstein, God is like the perfect physics textbook: The universe exists and operates according to absolutely deterministic principles. If you possess the proper and complete set of differential equations, you can plug in the values of all of the variables of reality and determine the exact state of the universe at any point in the future or past. The state of reality at any time is an unchangeable fact that derives with rigid necessity from the invariable laws of nature and the state of the universe immediately before. We do not know these laws and we cannot perfectly measure all of the quantities, but it is the sense of awe we experience in confronting the functioning of the completely structured universe that leads us to try to unravel the form of these governing principles through science. The religious impulse is the wonderment we have in facing the infiniteness of reality and the elegance of the harmonious simplicity that governs its intricate complexity. Science, the search for an understanding of the orderliness of reality, is the heart of Einstein's cosmic religion.

He found that universal order reflected in his theory of relativity. His first version, the special theory of relativity, unified mechanics (the theory of motion) with our best account of electricity, magnetism, and light, creating a new understanding of the fundamental concepts of space, time, motion, mass, and energy that allowed us to see the workings of reality with a coherence that had not previously existed. His extended version, the general theory of relativity, added gravitation in a way that transformed the combined space-time manifold into a pliable field that interacts with mass and energy throughout space and time. Einstein's use of the field concept unified and connected everything in a fashion that was entirely fixed by elegant equations. The cosmic religion was a guiding picture, a set of metaphysical assumptions about the underlying nature of the universe that was increasingly confirmed by Einstein's own advances.

But the determinism he inherited from Spinoza's Euclidean picture of the world was challenged by the reality he so adored. Quantum mechanics took physics beyond Einstein, whose most famous religious quip, "God does not play dice with the universe," was a reaction to quantum mechanics' inherently probabilistic nature.

Quantum mechanics is a theory that developed from the study of the atom. We had always assumed that the way things work in the world we see around us day-to-day must be the way they work everywhere. Einstein's theory of relativity showed that this isn't the case with things that are very big or very fast. Quantum mechanics showed that it isn't the case with things that are very small.

Physical theories start with what are known as "state variables." These are quantities whose values describe the state of some system. If I wanted to understand the behavior of a box of ping pong balls, I'd need to know their masses, positions, and velocities at a given time. What differentiates these properties from other properties—the color of the balls, the day of the week, or the maiden name of the mother of the person observing—is that they appear in what scientists call the "state equations," the mathematical rules that govern how the ping-pong balls move around the box. The determinism in Einstein's cosmic religion amounts to a belief that at every time, every state variable has a completely determined value and that there exists a set of absolute state equations according to which the entire universe as a system necessarily develops over time.

In quantum mechanics, the state equation is called the Schrödinger equation, named after the Austrian physicist Erwin Schrödinger. It has one state variable denoted by the Greek letter psi, ψ, and is usually referred to as the "wave function," because Schrödinger's equation has the form of the equation that governs the behavior of all sorts of waves. But what is ψ a measure of and what is doing the waving? There's not a good answer to either of these questions, so Einstein took to calling it the "psi-function" refusing to attribute the name "wave" to something that did not deserve it.

What is strange about quantum mechanics is that ψ is not a single measurable quantity. It has to be set out in terms of other measurable things, say, position or momentum. What is stranger still is that ψ is not a single value for this observable quantity, but a mathematical combination of every possible value of that observable quantity multiplied by a coefficient, a number between zero and one. In other words, if we measure ψ for some object in terms of position, then ψ is a combination of every possible location the object could occupy (yes, it could be an infinite list) with a fraction between zero and one attached to it. The Schrödinger equation describes how this mathematical combination evolves over time and, like Newton's and Einstein's equations, it is deterministic. If you give me the combination of coefficients and values for ψ at any one time and a description of the physical situation (that is, the potential energy the particle interacts with), I can completely and uniquely determine the coefficients and values for ψ at any other time by plugging it into the Schrödinger equation.

The problem is that we never observe something in this combined state, what physicists call its "superposed" state. Take one of these properties, location—where something is. Whenever you look for something, you find it to be somewhere, a particular place, not spread out in a cloud of probability all over. Everything we see is somewhere.

But according to quantum mechanics, when we are not looking at a thing, it does not have a single well-defined location. It is in a superposed state of being in every possible place it could be found. Let's send a photon, a particle of light, at a solid piece of metal with two slits cut into it, one on the left and one on the right. If we look at a screen on the other side of the piece of metal and do not check to see which of the slits the photon went through, the Schrödinger equation tells us that the photon is always in a combination of having gone through both the left and the right slits—even though it is impossible to break up a photon. Photons cannot be split in half, yet it turns out that, because of the alternating pattern of light and dark bands that can be seen on the screen, we know for a fact that the photons did go through both the left and the right slits simultaneously. This is because the pattern we see—the alternating stripes—can only be the result of two things (which act like waves) adding to and subtracting from each other.

But a really weird thing happens when we add a photon-detector into the system. The instant we start to check which slit the photon went through, we have two big changes. First, we will always see the photon with a definite location; it will always register as having gone through the left slit or the right slit. The superposed state of going through both slits simultaneously is never seen. The very second we begin to check to see which slit a photon goes through, the photon goes through one and only one slit. Second, the tell-tale stripes disappear from the screen, replaced with a series of random flashes—what we would expect if each slit sent out particles like a shotgun if the pellets came out one at a time.

The Schrödinger equation, with its bizarre combination of values, applies to a system with absolute certainty right up until the moment we look at it, at which point the system inexplicably leaves its superposed combination and instantaneously collapses into one of the observable values. So, the Schrödinger equation is a law of nature that only applies to the universe when we are not looking at it. This does not sit well with the tenets of Einstein's cosmic religion.

But it gets worse, because, when the photon collapses from the superposed combination into a single place (left or right slit), we have absolutely no way of knowing which one it will be. Even if you give me all the available information, the theory cannot determine right or left. We have an entirely random event governed by no deterministic rule. This is not to say that there aren't better and worse bets. The coefficients between zero and one that are attached to the values in the combination turn out to give us the probability that we will find it with that value of the observable property. We can set it up so the coefficients predict that half the time the photon will go through the left slit and half the time through the right slit. On any given observation, though, there is absolutely no way to know which it will be. This is the irreducible probability in quantum mechanics that so disturbed Einstein.

It isn't the use of probabilities themselves. Einstein was a master of the mathematics of chance. His doctoral dissertation examined the statistics of mixing, for example, allowing us to know how long we have to stir our coffee to get the cream evenly distributed through it. His contributions to statistical mechanics—the science that deals with the properties of collections with huge numbers of objects so that we can say something about the whole without having to calculate for each of the individual parts—are extremely impressive. In the sorts of cases Einstein dealt with, we have lots of things interacting with each other, so many that we cannot solve the equations governing the interactions because they become too complex. The best we can do are statistical approximations, something Einstein did as well as anyone.

In the cases Einstein prefers, we use probabilities because we simply cannot handle the calculations of the real underlying deterministic happenings. Flipping a coin, for example, is not really a random event. Suppose you know the weight of the coin, the shape of the coin, how hard it was flipped and where on the edge the pressure was exerted, the speed and direction of the wind, the height at which it was flipped, and so on. If you knew these seemingly infinite variables, then you could with absolute certainty determine the outcome. But there are so many of these variables, and we don't know the exact values of any of them when calling heads or tails, so it is because of our lack of knowledge that the event seems a matter of chance.

In quantum mechanics, however, the probability is irreducible; there is no mechanistic, deterministic calculation of any complexity underlying the collapse of the wave function. It simply *is* random—even if you know everything there is to be known about the system. Or so the advocates of quantum mechanics said, an interpretation of the mathematical symbols of the theory that Einstein vehemently opposed.

Schrödinger's equation was right as far as it went, Einstein argued, but it did not go far enough. Just as his equations in statistical mechanics were just pragmatic stand-ins for the real, more complex equations governing the actual reality of the situation, so, too, in quantum mechanics, Einstein contended, there must be hidden variables, new quantities to measure that if we only knew would tell us with certainty which slit we'd see the photon go through. Quantum mechanics had gotten every prediction correct. It had explained what it had set out to explain. But Einstein would not accept it on religious grounds, not as a Jew, but as a theological advocate of his cosmic religion. When Einstein said that "God does not play dice with the universe," he meant it about his Spinozan God.

Yet, while this opposition to quantum mechanics is in no way connected to his Jewish background, it is interesting that when Einstein reaches for a metaphor to make sense of his worries with the theory, it is a biblical reference. He wrote the following to Max Born:

Quantum mechanics demands serious attention. But an inner voice tells me that this is still not the true Jacob. The theory accomplishes a lot, but it scarcely brings us closer to the secret of the Old One. In any case, I am convinced that He does not play dice.[33]

While the objection itself to the standard interpretation of the new quantum theory that so bothered Einstein came from his cosmic religion, not from Judaism, its expression clearly combines the two.

AS AN ADHERENT OF HIS COSMIC RELIGION, Einstein did not consider himself to be an observant or theistic Jew, but he clearly was a cultural Jew. Einstein regularly referred to himself as a Jew when he used pronouns like "we" and "us" in addressing Jewish audiences.

He also had certain predilections common to members of the community. He loved Yiddish humor, for example. The physicist Abraham Pais recalled of his regular walks with Einstein in his later years:

One day I told Einstein a Jewish joke. Since he relished that, I began to save good ones I heard for a next occasion. As I told these stories, his face would change. Suddenly he would look much younger, almost like a naughty schoolboy. When the punch line came, he would let go with contented laughter, a memory I particularly cherish.[34]

One of the reasons Albert Einstein the scientist became Albert Einstein the celebrity is that he was not just a genius, but a *kibitzer*, a joker.

Further, he was in the mainstream of European Jewry at the time in being a Zionist. Einstein's Zionism developed out of a deep sense of care for the well-being of the global Jewish community, his community. His experiences from childhood through the Second World War convinced him of the Jewish community's fragility, and as he did as a boy in Catholic school, he considered himself a Jew in part because others, particularly the anti-Semites, did.

If we as Jews can learn anything from these politically sad times, it is the fact that destiny has bound us together, a fact which in these times of

> quiet and security, we so often easily and gladly forget. We are accustomed to lay too much emphasis on the differences that divide the Jews of different lands and different religious views. And we forget often that it is the concern of every Jew, when anywhere the Jew is hated and treated unjustly, when politicians with flexible consciences, set into motion against us old prejudices, originally religious, in order to concoct political schemes at our expense.[35]

When you are adrift in a life raft far out at sea with sharks circling, it makes no sense to deny that you have something in common with everyone else in the raft.

Einstein, like the four sons in the Haggadah, stresses the multifaceted nature of Judaism. "Jewish" has no set of necessary and sufficient conditions but is an umbrella term covering many different ways of being. Einstein is echoed by the Judaic scholar Michael Satlow, who argues that the fact that notions of Judaism have "no inherent meaning is actually a strength; they exist in a dynamic intertextual world in which Jews are able to link them to other practices, symbols, and texts to create transient and historically contingent meanings."[36]

In other words, just as Talmudic scholars bring out multiple meanings of biblical passages, so the entire concept of being Jewish is something that is—and must be—constantly redefined for the times. This ability to rethink and reshape Judaism is what has allowed it to survive and remain relevant to the living of real lives over its vast history. It is astounding that the same rituals could remain meaningful while being performed for thousands of years. Yet one could hardly say that the meaning they hold for contemporary Jews is the same as for their ancestors.

At the same time, Judaism would be an empty concept if there wasn't something underlying it, something binding Jews together. This was not something Einstein had left unconsidered. Indeed, he explicitly addresses it in an essay entitled "Is There a Jewish Point of View?" Einstein begins by contending that "in the philosophical sense, there is, in my opinion, no specifically Jewish point of view," but then sets out what he saw as the two elements that form the essential nature of Judaism.

First is an ethical stance toward the living.

> Judaism seems to me to be concerned almost exclusively with the moral attitude in life and to life. I look upon it as the essence of an attitude to life which is incarnate in the Jewish people rather than the essence of the laws laid down in the Torah and interpreted in the Talmud . . .
>
> The essence of that conception seems to me to lie in an affirmative attitude to the life of all creation. The life of the individual only has meaning insofar as it aids in making the life of every living thing nobler and more beautiful. Life is sacred, that is to say, it is the supreme value, to which all other values are subordinate. The hallowing of supra-individual life in its train a reverence for everything spiritual—a particularly characteristic feature of the Jewish tradition.
>
> . . . "serving God" was equated with "serving the living." The best of the Jewish people, especially the Prophets and Jesus, contented tirelessly for this.[37]

An essential element of Judaism according to Einstein is action springing from empathy. It is a moral stance based not on metaphysics or abstract principles but rather on relating to those living here and now in the world alongside of us.

Did Einstein himself fulfill this condition? Yes and no. Einstein spoke out and published in the name of peace and justice at times when it was not popular, indeed dangerous, to do so. In 1914, for example, he signed the "Manifesto to the Europeans," a document composed by the German pacifist Georg Freidrich Nicolai in response to the nationalistic "Manifesto to the Civilized World." This document was also called the "Manifesto of the Ninety-three" after the number of its famous signers, scientists, artists, and intellectuals who represented every area of German life and culture. The "Manifesto to the Civilized World" denied claims that Germany was guilty of provoking the First World War by invading a neutral Belgium, denied that Germany had committed atrocities there or that German troops had suffered military defeats, and contended that German militarism was a necessary defense of German culture. By contrast, the "Manifesto to the Europeans" stressed internationalism, considered by conservatives as treasonous, and Einstein received death threats for giving it his support.

Throughout his life, Einstein used his fame and social capital to seek relief for the oppressed. He wrote hundreds of affidavits for Jewish refugees seeking to escape Europe during the Second World War, so many that he joked that he was running his own immigration department. He spoke out frequently against discrimination—not only directed at Jews but also directed at the poor and Native Americans—against the rise of fascism in Spain, against the repressiveness of the Soviet regime, and against the treatment of African Americans in his adopted home. He had an ongoing correspondence with W. E. B. Du Bois[38] and wrote in his essay "The Negro Question":

> There is, however, a somber point in the social outlook of Americans. Their sense of equality and human dignity is mainly limited to men of white skins. Even among these there are prejudices of which I as a Jew am clearly conscious; but they are unimportant in comparison with the attitude of the "Whites" toward their fellow-citizens of darker complexion, particularly toward Negroes. The more I feel an American, the more this situation pains me. I can escape the feeling of complicity in it only by speaking out.[39]

Such advocacy on the part of the oppressed led J. Edgar Hoover to maintain a significant file on Einstein.[40]

But as much as Einstein embodied this Jewish ethos of "serving the living" in his public dealings, his private life was quite different. He contended that grappling with the mysteries of the universe required him to extricate himself "from the chains of the 'merely-personal.'"[41] This was Einstein-speak for neglecting his family. Einstein was a terrible husband and father, distant, moody, and at times cruel as his wife raised his two sons, one with psychological problems. As his first wife, Mileva, sank deeper and deeper into depression, Einstein became more and more distant, his sharp tongue rarely in check. He had many adulterous relationships while married to his second wife, Elsa, affairs he did not bother to keep secret from her. So, while he served the living in his public persona, privately, he often served his own needs and desires before those of the people who cared for him.

The second definitional quality Einstein ascribes to Judaism is awe.

> But the Jewish tradition also contains something else, something which finds splendid expression in many of the Psalms, namely, a sort of intoxicated joy and amazement at the beauty and grandeur of this world, of which man can form just a faint notion. This joy is the feeling from which true scientific research draws its spiritual sustenance, but which also seems to find expression in the songs of birds.[42]

By Einstein's own account, he was very much Jewish in this way his entire life.

> A wonder of such nature I experienced as a child of 4 or 5 years, when my father showed me a compass. That this needle behaved in such a determined way did not at all fit into the nature of events, which could find a place in the unconscious world of concepts (effect connected with direct "touch"). I can still remember—or at least believe I can remember—that this experience made a deep and lasting impression upon me. Something deeply hidden had to be behind things.[43]

As a young boy, there was wonder in the complexity and simplicity of the workings of the universe. This would lead Einstein to question and revise the deeply hidden nature of concepts assumed by even geniuses the likes of Isaac Newton "as being well known to all," notions like space, time, and motion. Einstein's joy and amazement would lead him to radically reformulate them in ways that made our understanding of the universe more amazing, more awesome, more beautiful. In this way, one cannot but call the theory of relativity Jewish science. But, then, one would also have to consider all great science to be Jewish science.

SO, IN THIS WEAKEST SENSE, is the theory of relativity "Jewish science?" Yes and no. Einstein satisfies certain definitions of "Jew" and not others. Einstein's case is as complex as the notion of "Jewish" itself. In a community that spans the globe and millennia, membership is a dynamic notion, defined by properties and relations within the community and between the community and the cultures within which it finds itself embedded. This idea of definitions being implicit in a frame of reference and between frames of reference would resurface in a very deep way within the theory of relativity's own mathematics. So while Einstein himself may or may not be considered Jewish, what about his theory? Is relativity Jewish?

Is Relativity Pregnant with Jewish Concepts?

The negative reaction we have to the claim that Einstein's theory of relativity is "Jewish science" is certainly conditioned by its association with Nazism. But phrases like "religious science" also bring immediately to mind views like contemporary Creationism. The idea of "religious science" as science that derives from, makes inextricable reference to, or is constrained by theological beliefs often connotes a closed-mindedness, an antiscientific bias in which some literalist interpretations of a given sacred scripture leads to a "come hell or high water" rejection of contrary positions no matter how much they are empirically supported.

In response, we've created a false dichotomy, a caricature in which working scientists become a combination of Joe Friday concerned with "only with the facts, ma'am" and Mister Spock engaged in purely logical reasoning from those facts, free from any influences from their times and culture. Meanwhile the effects of religion are held to be universally harmful to the scientific project, obscuring the rational and dispassionate inferences that follow from pure unadulterated observation.

But the caricature is flawed. While there are certainly significant cases in which the political power of organized religion has retarded our investigation of the natural world, science is a human project that occurs in a social and historical context of which religion is a part, indeed at certain places and certain times, a major part. We cannot understand why scientists did what they did when they did it without understanding the relationships between the science of an era and social institutions—religious institutions among them. Theology brings with it views of the universe and ways of thinking about puzzles. It would be absurd to think that these views and methods did not condition scientists to naturally think in certain ways.

In the history of geology, for example, the story of the Great Flood helped lead thinkers to form naturalistic hypotheses involving massive Earth-altering events as explanations for observed phenomena like the existence of rock strata and the finding of aquatic fossils at the tops of mountains. Catastrophism was the movement that began modern geological thought and traces its roots back to the biblical account.[1]

Both classical Chinese and Western astronomy had astrological aspirations, seeking to correlate celestial phenomena with human affairs, but they have taken fundamentally different approaches in studying the heavens. While the dipolar Yin and Yang of Taoism yield a world that is in constant flux from competing forces, the Western view, with its all-powerful lawgiver God, sought first and foremost to find strict rules and rational patterns that allow us to understand how he governs his Creation. Thus, Chinese astronomy focused almost exclusively on the rich irregularities that could be found in the motions of the heavens—the retrograde motions, eclipses, and the appearances of comets—and the Western astronomers focused on finding the clockwork mechanism by which the universe operates. Whether an astronomer was coming from a culture with one or two fundamental metaphysical forces controlling the universe determined what that scientist would look for when viewing the night sky and what sense he would make of it.[2]

Western physics, with its love of symmetry and conservation laws owes its fundamental approach to this theological orientation. That we see time not as circular, something implicit in pagan beliefs tied to the recurring seasons, but as linear, moving from a beginning possibly to a determined end, is biblical in origin. We think about science in the way that we do in part because we are heirs to a cultural legacy whose basic stance toward the universe derives from an Abrahamic sensibility. In this way, all contemporary Western science is to some degree religious science. The desire to understand the mind of an all-perfect and thereby super-rational Creator gave science its most foundational presuppositions that remain firmly in place long after these motivating factors ceased to be operative.

August Comte, one of the founding fathers of sociology, argued in the nineteenth century that each science proceeds through three phases: the religious, the metaphysical, and the scientific or positive.[3] The religious view posits an external intelligence whose will is responsible for observable phenomena. Why did that just happen? God did it. The move to second stage, the metaphysical, removes the anthropomorphic deity from the picture, replacing it with disembodied, magical forces. We go from "God pulls the apple to the ground" to "there is some motive desire on the part of the apple's matter to seek the ground as its resting place." The final mature phase, he argues, forms our understanding of the world based on bare correlations of observations absent any sense of motivation or cause. Science finds relational equations by which the universe operates. To ask why those rules are the rules is to ask for a story, a mythological explanation. It is the hallmark of immature prescientific thought.

But Comte was naive in thinking that, while the preexisting religious elements leave a residue on the later metaphysical approach, the ultimate move to the positive or scientific stage wipes away all marks from the previous phases and that no conceptual residue whatsoever is left from the historical religious roots of the science. The path to our current beliefs, even those supported with objective evidence, still bear the marks of their emergence. Elements of the cultural worldview, religious aspects included, can still be found. This is not to say that to do science today we have to buy into these beliefs or that scientists are secretly or unknowingly committing themselves to the existence of some God or other. Rather, it is to say that if we really want to understand why we believe what we presently believe, we cannot pretend that our beliefs do not have the biography they have. If not for the Judeo-Christian-Islamic[4] presuppositions, the scientific posture we now adopt would be very different.

But the influence of religion on science is not limited to this indirect sense of providing a basic stance toward the universe. In important cases in the history of science you can see the effects of individual authors' own religious viewpoints when you examine the details of their theories. To see whether the theory of relativity is Jewish science, let's compare the ways in which the theories of space and time that led up to it—those of René Descartes and Isaac Newton—derive from their own theological contexts and consider whether this could be the case for Einstein also.

RENÉ DESCARTES WAS A FRAIL AND SICKLY CHILD.[5] He suffered from a lung ailment, most likely tuberculosis, which kept him bedridden when most others his age were running around and playing. His mother died in childbirth while he was still an infant and his father, a powerful lawyer and politician, spent the better part of the year away from the family estate where Descartes and his brother Pierre were raised by their maternal grandmother and a nanny.

The nanny was an extremely religious Catholic woman, and she frequently spoke to Descartes of God and heaven, something that would keenly interest a sensitive boy who lost his mother. Descartes desperately wanted to believe, but he needed proof and even as a young child interrogated his caretaker, trying to gain the certainty he craved. The will to believe and the quest for certainty would remain competing goals that would shape the rest of his life. He was a devout Catholic, but, like many in the budding scientific community at that time, he was also an adherent of the Copernican heliocentric system. As such, one of the most formative events in Descartes's life happened to someone else.

Galileo Galilei became a Copernican evangelist just before the dictate of a group of theologians convened by Pope Paul V declared that the Aristotelian Earth-centered picture was dogmatically true. The Earth, the pope endorsed, does not move. Galileo begged—quite literally—to differ.[6]

Even though he was a professor of mathematics, Galileo had interests that were not purely abstract. He was a practical man to his core, seeking to advance his position in society by applying his great intellectual, practical, and social skills wherever they could have financial or personal rewards. Such opportunities arose in solving questions of martial concern such as projectile motion. Galileo was a part of the military-industrial complex long before it became complex.

So, when he heard of Dutch craftsmen creating a tube with lenses at either end that allowed distant objects to be seen as if near, he immediately realized the potential for its use by armies, navies, and merchant sailors—and the profit that could be made. He began to make and market them with all the showmanship of a master, allowing people to see ships coming into port from atop a tower in the Piazza San Marco in Venice.

But history changed when Galileo realized that he could also use his new instrument at night and became the first to see the heavens through a telescope. What he saw contradicted the Church's Aristotelian doctrine in many ways. Aristotle said that all bodies from the moon on out are comprised of a special, more perfect element called "aether" and that all such bodies, being more perfect, must be perfectly spherical. But Galileo looked at the moon and found mountains and craters. Aristotle said that all heavenly objects orbit the Earth. Galileo found moons orbiting Jupiter, not the Earth. He observed the phases of Venus, which failed to include a full phase, something that should be seen if Venus and the sun orbited the Earth as Aristotle claimed.

He collected all of this in his letters and writings, in which he argued that the heliocentric picture is the only one supported by the evidence—a claim that would land him in trouble. And in 1616, he was called before the Inquisition in Rome and forbidden from publishing anything on Copernicanism thereafter.

But he did not stop talking about it. In fact, he had a good friend, a cardinal named Maffeo Barberini, with whom he would take long walks and engage in spirited debates on the subject. When Gregory XV—Paul V's successor—died and the white smoke emerged, Barberini had become Urban VIII and the ever-enterprising Galileo saw his opening. Believing he now had protection, he requested permission to write a new book that gave a balanced treatment of the Church's Aristotelianism and the new Copernicanism.

The resulting work *Dialogue Concerning Two Chief World Systems* was anything but even-handed. It was a conversation between three figures: Sagredo, who does not know what to believe, Salviati the Copernican, and Simplicio the Aristotelian.[7] Galileo claimed that the name "Simplicio" refers to the Neoplatonic philosopher Simplicius, but the insult was far too obvious. Coupled with the fact that the character could easily be interpreted as a caricature of the pope, Galileo once again found himself before the Inquisition in 1637. This time, he was not let off so easily. Placed under house arrest for the rest of his life and forbidden from working on astronomy, the Church decided to make an example of Galileo.

The scientific community clearly noted the message, but Descartes took precautions to a level beyond most. He realized that despite the depth and authenticity of his personal faith, his views could still land him someplace he did not want to be, so like the Earth, he too would have to constantly move and he took refuge in Protestant Holland, where he would be a bit freer to work on thoughts that might not be in line with Catholic orthodoxy. But he worried that merely relocating to the Netherlands would still be insufficient for personal safety and so Descartes changed his residence every few months and put false return addresses on most correspondences so that surprise visitors would never find him.

This caution extended to his intellectual work, and when coupled with his desire to adhere to the word of the pope, he was led to consider ways in which Copernicus's heliocentric model could be made consistent with the Catholic Aristotelian point of view. Where Galileo challenged this position head on, arguing that observations make it clear that the Earth does move around the sun, Descartes, unwilling to embrace heresy, tried to finesse the problem. He thought that by producing a larger theoretical framework through which to understand the motion of the Earth, he could have it both ways—maintaining a stationary Earth but keeping the Copernican picture of modern science.

Descartes begins his book *Principles of Philosophy* by arguing that the defining characteristic of space is extension.[8] If we can say that there is a real distance between the Earth and the moon, then that measurement has to be the property of some real thing with that particular length, namely, the space between the Earth and moon. Since we can only measure real things (it is nonsense to ask "How tall is Santa Claus?"), space must be seen as a real thing since we can measure lengths, areas, and volumes of it. Space itself, therefore, becomes an object. The concept of a vacuum, of a completely empty region of space, becomes self-contradictory, because you can't have a space empty of space. Space, in Descartes's view, is the substance that fills the universe.

Further, when something moves, Descartes argues, there must be something moving it, something in physical contact with it that causes it to change location. Descartes's mechanistic worldview demanded that all motion be the result of pushes, pulls, twists, and things bumping or rubbing up against other things. The universe is like a watch where meshed gears pull at each other, the tension moving one and the connectedness thereby driving the rest. If space is a thing, it could mechanically interact with the stuff embedded within it. Just as a leaf floats downstream pulled by the current of the water, so, too, we could make sense of gravitational motion as the space-filling space dragging along material objects.

But so much of the motion we observe in the universe is circular, so Descartes posited that each massive object creates a vortex in space around it, a whirlpool of space stuff that then pulls other objects in a circular orbit about it. Think of a bathtub after you've pulled the drain plug, except that the amount of water somehow stays constant. A rubber duck floating in the tub would circle the drain just as a planet circles the sun, but if it gets too close to the drain it would get pulled straight down, just the way a ball dropped from a tower behaves. Thus with Descartes's vortex theory of gravitation we have a mechanical explanation for the motions of heavenly and earthbound bodies.

But how do we square this with the papal dictate? The Earth cannot move. Period. While Descartes held out some hope that theological whims of the pope's inner circle might reverse itself on Copernicus sooner or later, the philosopher sought some means of squaring his Copernicanism with the seemingly mutually exclusive doctrine of his faith. How to do this? The answer requires extreme cleverness.

Consider the statement "the Earth does not move." What do we really mean by this? Suppose you are on the up escalator at the mall. Are you moving? From the point of view of someone also riding the escalator, no, you are standing still. From the perspective of someone standing on the first level looking at you from casual wear, yes, you are moving up and over. Who is right? Are you *really* moving or not?

To answer this "really" question, we need a definition of motion. We know we can speak of motion relative to something else, but if all motion is merely relative, then there is no "really," no fact of the matter. This is obviously unsatisfactory when it comes to the papal decree, which clearly intended to say something deep and meaningful about the actual state of the Earth, not just that it is at rest with respect to something or other, but absolutely, truly, and completely at rest. If we want to say the Earth *really* doesn't move, then it must be at rest with respect to some absolute thing that—by definition—is also not moving. What could this be? The pope's theologians didn't say.

Descartes argues that the only thing absolute enough to do the job is space itself. An object is moving if it is changing the space it is surrounded by. If I drive from Washington to New York, it must be true that I have moved because the space stuff around me in D.C. is not the same space stuff that surrounds me in Manhattan.

But—and here's the trick—the Earth has an atmosphere. It carries its surrounding air around with it as it rotates and revolves. Perhaps it's not just the air that gets dragged along with the Earth, but the underlying space as well. Think again of our ever-full bathtub without the drain plug, only now consider a water balloon floating in it. The balloon is pulled by the water and circles around the drain. Now, put a marble inside the balloon. Since it remains surrounded at all times by the same water inside the balloon, the marble would not be moving according to Descartes's definition of motion, even though it too is circling the drain.

If we take the Earth to create a vortex in space that keeps the same exact bits of space around it, then by Descartes's definition of motion, the Earth does not move, even as the sun's larger vortex carries the Earth's whirling region of space around it. Since the absolute motion of an object is defined as motion relative to the space around it, it may be true that the space around the Earth orbits the sun, but, because the Earth's vortex keeps it forever bathed in the same space-filling space, it is entirely stationary. Hence, we have both Aristotle's nonmoving Earth *and* Copernicus's revolving Earth at the same time. Descartes gives us a Catholic theory of gravitation that is scientific in its goal to fit theory to fact while adhering to the word of the pope. Darn clever.

ISAAC NEWTON NEVER MET HIS FATHER, who died before his birth.[9] The desire to know and please his father who art in heaven shaped Newton's mind. Newton's mother soon remarried. As her new husband had no desire to raise some other man's child, Isaac was sent to live with his grandmother until the new husband died leaving her financially secure. The bond between Newton and his mother was extremely close. Indeed, Newton had virtually no serious emotional relationship with any other person his entire life and never felt at ease in the company of anyone else, being extremely shy and socially inept.

Newton was sent away to grammar school and lived with an apothecary. Through him, Newton learned the basic skills of a scientist, becoming obsessed with alchemy and the workings of mechanical systems. Without friends, he took to building complex toys like model windmills and to performing experiments.

From the potions of his youth, he would acquire a sense that there were principles at work, mechanisms unseen, whose mysteries were those of Creation and its Creator. Throughout his entire life, Newton worked at both theology and physics with a dedication that led to a virtual disregard for bodily well-being. The two studies were, to his mind, entirely connected.

Newton believed that his results were known to the ancients, knowledge that had been lost. The Greeks, he contended, had been in contact with the ancient Jews, so that Moses, Pythagoras, and the

Egyptian alchemist god Hermes Trismegistus[10] contemporaneously contributed to a web of knowledge in the classical world. The synthesis of the Hebraic access to the Divine with the experimental results of the Egyptian practitioners of natural magic, when united with the Greek love of structured reasoning, provided the ancients with wisdom, Newton thought, that we were only beginning to rediscover, insights into truth and reality that had long been lost to humanity.

As a student at Cambridge, Newton was introduced to the best science and mathematics of the age and that meant Descartes. Newton became an expert in the Cartesian system and one can only look at aspects of his mature scientific work as extensions and refutations of it. Just as Descartes's Catholicism was an active part of the worldview that gave rise to his theory of gravitation, so too we find corresponding Protestant influences in Newton.

Newton's theory of the universe has two parts: a theory of mechanics and a theory of gravitation. With a nod to his reverence for the ancient Greeks, Newton structured his master work, *Mathematical Principles of Natural Philosophy* (generally referred to by the first word of its original Latin title, the *Principia*),[11] like Euclid's *Elements,* beginning with definitions and positing the three laws of motion as axioms, then working out cases as if theorems. Newton defines concepts such as mass, impressed force, and centripetal force, then supplies an addendum, a Scholium, in which he discusses four notions he will not define because they are "well known to all": time, space, place, and motion.

Newton respected Descartes's work but was deeply disturbed by two aspects. First, in Descartes's mechanical philosophy, the universe is a big clock in which things are only moved by contact with other things. This takes God out of the universe. God creates the world but then is absent from its day-to-day workings. This alienation of the Creator from his Creation, the absence of an ever-present God within the world, was abhorrent to Newton's theological posture. Descartes's absent God violates the Protestants' central notion of a personal and present God.

Second, Newton found the Cartesian notion of a movable space to be a perversion of the concept. Things move *in* space, space itself does not move, after all, what would it be moving in? The cleverness of the Cartesian attempt to match physical theory with papal decree did not amuse the Protestant Newton, who saw the artificiality of the entire approach as a mark of undeniable error. Newton is clear that with respect to time, space, place, and motion we must distinguish the "absolute, true, and mathematical" meanings of the term—his own sense—from their "common, apparent, and relative" counterparts—Descartes's usage.

Absolute space is "fixed and immovable." Unlike the Cartesian space that flows in vortices and carries objects along, Newtonian space is nailed down and inert, neither moving nor affecting the matter it contains. Similarly, absolute time "flows equably without relation to anything external" and absolute place is a position, an address in absolute space. And if place and time are absolute, then so is motion. Descartes's physics is based on what Newton sees as flawed definitions that give us merely apparent motion, not true, absolute, and mathematical motion, the sense of motion that ought to be employed in science. When we switch to absolute space, however, the Cartesian trick that allowed the Copernican view to remain consistent with the pope's decree is ruled out.

But like Descartes, Newton's basic notions were not just intended to account for observable phenomena but were also based on his core theology. Understanding the workings of the world is to read the mind of God.

> This most beautiful system of the sun, planets, and comets, could only proceed from the council and dominion of an intelligent and powerful Being. And if the fixed stars are the centres of other like systems, these, being formed by the like wise council, must all be subject to the dominion of One; especially since the light of the fixed stars is of the same nature with the light of the sun.[12]

The Reverend Richard Bentley preached that the Newtonian picture of the world demonstrates the design of an intelligent Creator. Newton was pleased with the sentiment, writing to Bentley:

When I wrote my Treatise about our System, I had an Eye upon such Principles as might work with considering Men, for the Belief of a Deity, and nothing can rejoice me more than to find it useful for that Purpose.[13]

But it is not just that the orderliness of the universe demonstrates the fingerprints of Divinity, according to Newton. God is literally in the Creation. From Query 28 of his work *Opticks*:

Is not the Sensory of Animals that place to which the sensitive Substance is present, and into which the sensible Species of Things are carried through the Nerves and Brain, that there they may be perceived by their immediate presence to that Substance? And these things being rightly dispatch'd, does it not appear from Phæomena that there is a Being incorporeal, living, intelligent, omnipresent, who in infinite Space, as it were in his Sensory, sees the things themselves intimately, and thoroughly perceives them, and comprehends them wholly by their immediate presence to himself?[14]

And from the *Principia*:

This Being governs all things, not as the soul of the world, but as Lord over all; and on account of his dominion he is wont to be called *Lord God* . . . or *Universal Ruler* . . . he governs all things, and knows all things that are or can be done. . . . He endures forever, and is everywhere present; and by existing always and everywhere, He constitutes duration and space.[15]

Space is God's sense organ, his eyes and skin. The reason God can be omniscient is that he literally is everywhere. Space is a part of God and so he knows instantly of all events. Since God is infinite, space is infinite. Since God is unchanging, space is unchanging. Since God is absolute, so too space must be absolute.

Within this fixed and immovable space are masses that interact with one another. Newton sets out three rules by which they do so.

These laws, known universally as Newton's laws of motion, have similar, but less developed counterparts, in Cartesian physics. The first law describes how things move when there is no external force applied to them. Where for Aristotle an object's natural motion is rest in its natural place, for Newton it is motion in a straight line at a constant speed. Aristotle contended that a force was necessary for any motion. Newton disagrees, asserting that a force is necessary to alter an object's current state of motion.

The second law describes what happens when a force is applied to an object. It accelerates proportionally to the amount of mass contained within it. Push something and it speeds up, slows down, or changes direction. The more massive a thing is, the more force needed to divert it from its trajectory.

If the first law describes the behavior of an object with no net force applied to it and the second what happens when there is a force exerted on an object, then a third law is needed to describe what happens to the thing that applied the force. It also experiences a force, one equal in amount, but opposite in direction.

With these three simple rules, we can account for the mechanical interactions of ping-pong balls. But the universe is more than ping-pong balls. It has planets and stars and for this we need one more piece, gravitation. Newton's law of universal gravitation asserts that there is an attractive force pulling every massive object toward every other massive object that is proportional to the product of the masses and inversely proportional to the square of the distance between their centers of mass. This means that if you double the amount of stuff in something, you double its gravitational pull. Take it twice as far away and it has one-fourth the pull.

It is called "universal" because, contrary to Aristotle, it is the same force that operates on Earth and in the cosmos. This is the insight that Newton supposedly gleaned from the falling apple, a story that is almost certainly apocryphal. Newton himself penned the tale to try to guarantee his priority for the discovery by placing his epiphany in the year 1665, when Cambridge had shut down because of the plague. He was supposedly off by himself on the family farm, Woolsthorpe

Manor, sitting beneath an apple tree looking up at the moon when his eye was distracted by a falling apple. This sparked the realization that the force pulling the apple must be the same as the one that keeps the moon in its orbit. Aristotle argues that there are distinct terrestrial and astronomical physics. On Earth the only possible type of movement is straight-line motion and in the heavens, which are perfect, the only possible type of motion is a circular path, including the epicycles—circles within circles—of Ptolemy. Newton's theory is different in that offered a unified account of both. But also one with an important property—it mandates imperfect elliptical orbits.

Before we look at how Newton's theories allowed for movement in space that was not a perfect circle, let's look back at other astronomers who challenged the Aristotelian system. Johannes Kepler was the mathematical assistant to Dutch astronomer Tycho Brahe, the greatest of the pretelescopic astronomers who compiled incredible amounts of the finest observations of the heavens that had yet been collected. So jealously guarded was this data, that he would not let Kepler take any of it home to work on and only allowed him to see isolated bits and pieces at the office. When Brahe died, however, it all went to Kepler, who decided to work out once and for all the shape of a planet's orbit, everyone knowing full well that it could not be a simple circle that would make things so tidy. Others, most notably the Greek astronomer Ptolemy, who was tied to the Aristotelian circularity, added epicycles to the orbit of heavenly bodies—circles on circles—and ecliptics—centers of rotation that were not the centers of the orbit so that planets flopped about like flat tires. Kepler wanted to eliminate these artificialities and, through painstaking work, discovered that the orbits are oval shaped, not circular as Descartes's vortex theory described.

But why elliptical orbits? What caused the planets to move in such odd paths? Kepler had no clue. It became one of the great scientific problems of the age. It was Newton's equation for gravitational force that finally answered the question, something that Newton and his supporters cited as reason to believe his account over Descartes's that gave circular orbits.

Descartes was wrong on the facts, Newton contended, but he was also wrong on the philosophical and theological foundations. Descartes held that the universe is comprised of two fundamentally different kinds of stuff—minds and bodies. The world of material bodies was governed by mechanical laws that required things to touch in order to have any effect. God, of course, could violate these laws if he so chose, but a perfect God, the Cartesians argued, would find no reason to do so in a Creation that partakes of his perfection.

But, as the Protestant Newton detested removing God from the workings of the universe, God was not absent from the world. So Newton had little concern for the Cartesians' furious objection that he provided no explanation of what causes the attraction across vast distances of empty space. "What was the mechanism behind gravitation?" they howled. Newton's response was simply, "Hypotheses non fingo," I frame no hypotheses.

> But hitherto I have not been able to discover the cause of those properties of gravity from the phenomena, and I frame no hypotheses; for whatever is not deduced from the phenomena is to be called an hypothesis; and hypotheses, whether metaphysical or physical, whether of occult qualities or mechanical, have no place in experimental philosophy.[16]

I have no idea *why* it happens, Newton contended, all I can do is mathematically describe *how* it happens.

Of course, this was a little disingenuous. He did have some idea. Writing again to Bentley, he argues that there are two possibilities: gravitation is internal to the bodies or externally imposed on them. It could be that attracting each other is just what massive bodies do, that having the magical power to gravitationally pull other masses across the great vacuum of space is just a part of what it means to be an object. But he explicitly denied this in another letter to Bentley:

> You sometimes speak of Gravity as essential and inherent to Matter. Pray do not ascribe that Notion to me; for the Cause of Gravity is what I do not pretend to know, and therefore would take more Time to consider it.[17]

So, if it is not in the bodies, then

> Gravity must be caused by an Agent acting consistently according to certain Laws; but whether this Agent be material or immaterial, I have left to the Consideration of my Readers.[18]

Gravity is mediated by some Agent who may be immaterial and is capable of acting on all masses simultaneously and continuously throughout all of infinite space. Now, who might that be, hmmm?

So Newton's theory, the one we learned in high school, the one that allowed us to figure out how fast a block of mass m moved after t seconds when placed on a frictionless plane inclined at an angle θ, derives in part from Newton's views of the Divine. We don't need to accept the theological underpinnings to use the theory, of course, but Newton's own understanding of it and his reasons for creating it as he did are inextricably connected to his religious worldview. In its content and in his own mind, Newton's science is religious science.

ALBERT EINSTEIN ATTENDED THE POLYTECHNIC UNIVERSITY in Zurich, the Eidgenössiche Technische Hochschule, known by its initials ETH. The physics department was led by Heinrich Weber: powerful, respected, and scientifically conservative. Einstein's interests were in the burgeoning theory of the electron and new approaches to the nature of light and heat. He had his own ideas about what to study and what experiments he wanted to carry out. But it was Weber's laboratory, and Weber had little patience for these trendy fashions, especially when it came to Einstein, whose antiauthoritarian attitude was coupled with a youthful arrogance and a propensity for skipping class. Indeed, Einstein likely would not have even graduated had it not been for his ever-diligent friend Marcel Grossman, Einstein's guardian angel, who attended class religiously and gave him access to his notes.

Personality clashes often end badly when there is an imbalance of power. After graduation, Einstein was passed over for a research assistantship at the ETH. Not too surprising, given his run-ins with

Weber and his less than committed work habits. But then he found himself uniformly rejected from every other position he applied for. No physicist would bring on a graduate of the ETH without getting some sense of him from Weber. Einstein sent letters to every physicist in Europe, but there was absolutely no interest from any quarter. He was shut out of the world of physics. He had no job and no prospects. He would, however, have a child.

Smart, independent, and sarcastic, Mileva Marić was the only woman in his physics classes and Einstein fell for her, hard. He had passed his exit exams for graduation, but she had not. At a rendezvous in Italy, Mileva became pregnant and moved back to her family in Serbia. Their child, Lieserl, was hidden, likely adopted by relatives as was common for children born out of wedlock at the time.

Einstein needed money. He needed work. Against his parent's wishes, he and Mileva intended to marry, but he could not afford the married life. He tried to make ends meet tutoring, but it wasn't enough. Einstein was desperate when his friend Grossman once again saved him. Grossman's father had connections, and Einstein was hired as a patent clerk in Switzerland, a stable, decent paying, civil service job that would be the key to a settled life.

Moving to Bern, he found a comfortable apartment. Mileva could not join him because they were not married, and having an illegitimate child would cause talk that could cost him his job. They wrote love letters and dreamed of the day they could be together as a real family. Then Lieserl contracted an illness, possibly scarlet fever, and while there is no record, from their letters it is believed that she died. Einstein never met his daughter. Mileva had retaken the university exit exam and again failed. She would never have the credential she worked so hard for.

The tragic death of Lieserl allowed them to marry and soon they had a son, Hans Albert, the three of them living in the small apartment. The marriage, however, was not to be a happy one. Einstein soon grew cold, rejecting the emotional needs of his family, what he termed the "merely-personal." Mileva, already angry, grew more and more resentful, which only made Einstein more and more aloof. He retreated into his physics. Before they were married, Einstein wrote letter after letter about the physics they would do together. Yet, now it was the wall he used to separate himself from her, using it to remain isolated from his contentious life.

In this dysfunctional personal situation, Einstein wrote four papers. One was his doctoral dissertation on the physics of mixing. Another was a treatment of Brownian motion—the strange zigzag paths of tiny particles of dust trapped in a liquid—and it helped lead to the widespread acceptance of the existence of atoms. A third explained the photoelectric effect in which electrons were emitted from a metal exposed to ultraviolet light. This helped begin the study of quantum mechanics and was the basis of his being awarded the Nobel Prize in 1921. The other paper, called "On the Electrodynamics of Moving Bodies," introduced the theory of relativity to the world. Einstein's *annus mirabilis*, his miracle year, brought him professional success during a period of personal crisis.

THE THEORY OF RELATIVITY WAS Einstein's ultimate rebellion. In physics, there was no greater authority than Isaac Newton and here was Einstein, a mere patent clerk with no Ph.D., not even an apprentice scientist, who had the *chutzpah* to claim to overthrow the most well-entrenched theory in history. But was it Jewish?

No. We know what he was reading, what questions intrigued him, and what figures were coloring his thought as he devised this theory. None of these influences were Jewish.[19]

Einstein's physical thought was deeply influenced by Heinrich Antoon Lorentz, a Dutch physicist who was concerned with the properties of the newly discovered electron and on the nature and properties of light. We know from Einstein's correspondences that he read Lorentz's papers and admired them immensely.

Light was thought to be a wave because of well-documented phenomena like interference in which light added and subtracted in the same sort of ways that we see with water waves coming in and out on a beach. Such behavior only made sense in terms of waves, not as particles. But waves need something to do the waving, a medium of propagation. No water, no water waves. What then about light, which travels to us across the vast distances of empty space from the farthest stars? The fact that we do see the glowing of stars implies that even empty space contains some medium capable of supporting light waves. It was given the name "luminiferous aether" to distinguish it from Aristotle's ever-present aether, and one of the main problems facing physicists at the time was determining its nature and properties.

The argument about the nature of the aether was a new version of the old debate between Descartes and Newton over movable and immovable space. Descartes's space flowed and dragged around objects, while Newton's space was fixed, immovable, and had no effect on the things in it. The same two options were considered for the luminiferous aether: one position held that moving objects cause disturbances in the luminiferous aether, like a fan moves the air, while the other view argued that objects moved through it like a butterfly net, leaving it unaffected.

Armand-Hippolyte Fizeau, a French physicist, attempted to experimentally settle the matter by sending water through a U-shaped pipe and measuring the speed of light sent through the water both with the current and against it. His results seemed to show that the luminiferous aether was dragged along, a "Cartesian" result. American physicists A. A. Michelson and Edward Morley, using a series of mirrors that allowed a beam of light to be split and recombined, tried to measure this dragging but found that there was no detectable "aether wind." According to their experiment, the aether had to be stationary, a "Newtonian" result. But the luminiferous aether couldn't both move and not move. Something had to give.

Lorentz approached the Michelson-Morley experiment with concern. He wrote, "This experiment has been puzzling me for a long time, and in the end I have been able to think of only one means of reconciling its results."[20] That "one means" was to figure what sort of effect moving electrical charges would have to have on the aether that would account for Michelson and Morley's puzzling finding. It turns out that we could account for the effects if the aether and everything in it contracted, squished, in and only in the direction of motion. Surely it didn't, Lorentz thought, and he hoped that there were other, less bizarre explanations, but the "Lorentz contraction" did the trick mathematically. It was a trick that caught Einstein's eye.

In addition to Lorentz and the luminiferous aether, Einstein was thinking about the nature of science. With his friends Michele Besso, Conrad Habicht, and Maurice Slovine, the tongue-in-cheek self-titled "Olympia Academy" was formed to discuss the works of great thinkers about the scientific method while eating sausages.

We know that they read Ernst Mach, a famous physicist who also wrote on the philosophy of physics. His view, positivism, contended that one ought to grant real existence only to observations, measurements, experiences, and sensations. It is the observations of things that are real, not the things themselves. Saying that material objects have a reality separate from our perception of them, Mach argues, is primitive, metaphysical, and speculative at best.

Chief among the things that Mach sought to get rid of were the Newtonian notions of absolute space and absolute time. He believed that we can observe relative locations and relative speeds. Recall that for Newton these were merely common and apparent notions to be replaced with the more noble—and theologically based—absolute, true, and mathematical versions. This led to the metaphysical reality of space and time independent of the things in them. Newton, Mach argues, is overstepping his bounds in foisting these metaphysical objects on us.

> It is scarcely necessary to remark that in the reflections here presented Newton has again acted contrary to his expressed intention to only to investigate *actual facts.* No one is competent to predicate things about absolute space and absolute motion; they are pure things of thought, pure mental constructs, that cannot be produced in experience. All our principles of mechanics are, as we have shown in detail, experimental knowledge concerning the relative positions and motions of bodies. Even in the provinces in which they are now recognized as valid, they could not, and were not, admitted without previously being subjected to experimental tests. No one is warranted in extending these principles beyond the boundaries of experience. In fact, such an extension is meaningless, as no one possesses the requisite knowledge to make use of it.[21]

Absolute space, time, and motion, like all other metaphysical claims (including the existence of God), are meaningless.

Of course, there were pesky facts that made this move tough for Mach. By doing away with absolute space, you get rid of the absolute reference frame. All motion becomes relative. When I am on the escalator looking at you on the floor below, there is now no answer to the "really" part of the question of which one of us is really moving.

Newton's ace in the hole here was to point out that we *can* tell which one of us is really moving when we move from cases of constant velocity to acceleration. Let's go from the mall to the amusement park and take me off the steady moving escalator and put me on a roller coaster. Again, our movement seems to be symmetrical. From the ground, you see me rising slowing, dipping and speeding up, turning and looping. Looking back at you, I see you doing everything in an equal, but opposite, way. It seems completely relative to say which one of us is moving.

Until, that is, I get off. There's only one of us who is turning green and feeling like my part of the bag of popcorn we shared may make a second, less appetizing appearance. It was my stomach and mine alone that experienced the forces from the acceleration, and it is the appearance of these forces and their effects, Newton argues, that allows us to distinguish between our absolute states of motion. I feel sick because I was in motion relative to absolute space. You feel superior with that smug look on your face because you were just watching, stationary to the fixed space. Absolute space and time gives us absolute motion, which can explain the asymmetry in motion sickness.

Mach thinks he can account for this difference without an absolute space by pointing to the stars. To him, space itself has no independent existence, but, no matter where you are, you will always be surrounded by observable stars that are extremely big and heavy and thereby have combined gravitational pulls that will have real effects, say, on my stomach. It is not that I was in motion relative to absolute space that caused me to feel queasy, it's that I moved relative to the stars and the subtle changes in gravity generated the forces responsible for my condition. In this way, we get what Newton derived from absolute space without the metaphysical commitments.

The Olympia Academy also closely read Henri Poincaré, a mathematician, physicist, geologist, and philosopher who is widely considered the smartest person of the generation before Einstein.[22] He was the last mathematician to make substantive contributions across the entire spectrum of the field. He was not merely a master technician but also an acclaimed intellectual whose popular writings on science and the philosophical foundations of science were widely read.

Like Mach, he was troubled by the basis of Newton's theory and felt the need to backfill more solid philosophical underpinnings. Newton had claimed, for example, that his laws of motion were the result of generalizations from observations, regularities that emerge organically from the data. But consider the first law of motion, that an object not subject to any net force will move in a straight line at a constant speed. How did this get derived from observation, when it is impossible to have an object subject to no net force? In a universe filled with things, all of which are pulling on each other gravitationally, Newton's own law of universal gravitation entails that no object of the sort described in the first law of motion could even exist to be observed. So Newton gives us a law of nature that can't actually apply to anything in nature and then contends that he got it through observation. It seems like claiming to have used empirical evidence to support a thesis about the biology of unicorns.

But surely it is meaningful. We just need to think of it differently. Don't consider this, or any of Newton's other laws, to be empirical statements about how the world is. Rather, take it to be a definition. Poincaré holds Newton's genius to be the insight that there are definitions needing to be laid down as preconditions for the possibility of observing.

Poincaré advocates a variation of the view of human knowledge proposed by Immanuel Kant.[23] It is naive, Kant argues, to think that observations just appear in our minds like fully constituted images sent from the senses. What we get from our eyes are not images of things, but a two-dimensional blur of undifferentiated colors. The mind takes that raw manifold of perception and shapes it, molds it, groups together blotches of color, creating the images of the objects we see. There is an active place for consciousness in helping to construct our experiences. Observation is a combination of matter coming in from the senses and order imposed on sensations by rational categories built into the mind. Kant held that this world-shaping faculty is innate. All humans have it, and we cannot but see the world the way we do because we all have the same set of basic categories.

This is where Poincaré differs. He takes Kant's central insight to be true, that there are preconditions to the possibility of experience, that there are notions that need to be in place before we talk about empirical matters, but, unlike Kant, Poincaré thinks that we can alter them. Kant thought that Euclidean geometry was one of these basic set of notions that we use to create the world and that the human mind could not think geometrically in a way that differed from that which follows from Euclid's postulates. But as a mathematician at the turn of the twentieth century who made contributions to geometry (indeed, he created topology, the study of what sits underneath all the different geometries), Poincaré knew that there were alternative systems that differed only in their basic notions.

Where most thinkers considered these systems to be competitors—which one is the true, Euclidean or non-Euclidean geometry?—Poincaré saw it differently. He saw them as different languages. Just as we can say that it is snowing out in French or in German and have it mean the same thing, just as we can read a thermometer in degrees Fahrenheit or Celsius to express the same temperature, so too we could explain the geometrical arrangement of things in Euclidean or non-Euclidean terms. Since geometry is a property of space, what we say about space becomes a matter of freely chosen, inter-translatable, linguistic conventions, not metaphysical truths.[24]

The same is true of time. Unlike the Newtonian view in which God's Rolex determines the absolute duration between events, Poincaré argues that there are assumptions implicit in our understanding that are so basic that we don't even realize that we're making them. If I want to know how long it takes you to run from here to that tree, I could look at my watch when I see you leave and again when I see you get to the tree. But how do I know whether my watch is right?

> In fact, the best chronometers must be corrected from time to time, and the corrections are made by the aid of astronomical observations; arrangements are made so that the sidereal clock marks the same hour when the same star passes the meridian. In other words, it is the sidereal day, that is the duration of the rotation of the earth which is the constant unit of time. It is supposed, by a new definition substituted for that based on the beats of the pendulum, that two complete rotations of the earth about its axis have the same duration. However, the astronomers are still not content with this definition.[25]

Newton argued that physicists make use of absolute time, not merely relative time, but how can they when we have no access to God's Rolex, only our untrustworthy Timexes?

Yet, we need some standard if we are to talk about time in our scientific equations. We need a definition, but like all definitions we are free to choose what we mean. We can pick any naturally recurring event, declare it to be constant, and set (and regularly reset) our watches to it.

But it gets even more complicated. When I look at my watch as you start to run, I am assuming that the ticking of the second hand is simultaneous with your first step. Of course, it isn't. What is simultaneous are my perception of the movement of the second hand and my perception of your starting to run. It is not the events themselves that happen at the same time, but my experiences of them. We have no access to the actual time associated with the events themselves. How can I ever know whether two distant events are really simultaneous?

Poincaré's response is that this is not a "really" question, but yet another example of the need for a definition. We *define* simultaneity for distant events, we do not discover it. Crucial aspects of both space and time are therefore neither empirical facts of the world nor metaphysical necessities that flow from theological considerations, rather they are nothing but arbitrary conventions, definitions that we could make in any of a number of ways.

We find these influences clearly shaping Einstein's thought in his paper "On the Electrodynamics of Moving Bodies,"[26] but we also find classic Einstein. As he so often does, he leads off with a thought experiment. Take a coil of wire and a magnet. Move the magnet back and forth inside the coil, generating an electric current. Now hold the magnet still and move the coil back and forth over it at the same rate. The result? Same current. All that matters is that one is moving with respect to the other. The problem is that our best theory of electricity and magnetism gives different explanations depending on which one is really moving. The "really" question here is the result of invoking the luminiferous aether as an absolute reference frame.

Einstein does to the luminiferous aether what Mach did to Newton's absolute space: he eliminates it as needless metaphysical baggage. Following Mach's line of argument, all that matters is the relative motion between the magnet and coil. Hence, there is no "really" question about which is moving and, therefore, no need for the luminiferous aether at all, since we have no need to determine which is really moving and which is really at rest.

Einstein takes Mach's ideas and makes them one of the two pillars upon which to build the theory of relativity. The "principle of relativity" asserts that the laws of physics should be the same for all reference frames that move smoothly with respect to each other. Acceleration causes weird forces to pop up, but when any two observers moving in a straight line at a constant speed look at the world, they darn well better see it operating according to the same exact rules.

Paired with this is the "principle of the constancy of the speed of light," which is a significant aspect of Lorentz's treatment. If I walk at two miles an hour up the escalator that is itself going up at one mile an hour, I should be moving up at three miles per hour. That's what Newton said. But light doesn't work like that. If I am shining the light from a flashlight in your eyes, you see it coming at you at the speed of light. It turns out that the speed at which you see the light approaching you does not change at all if I am standing still, running toward you at a hundred miles per hour, or away from you at a hundred miles per hour. Newton's laws can't handle that.

It was Lorentz who began to figure out how to account for this odd property and Einstein who finished the work. Step one is getting rid of the luminiferous aether, but then we would need to completely rethink the nature of space. But as goes space, so goes time.

The second step is to follow Poincaré in seeing the need to set down a definition of simultaneity before beginning to talk about time as an empirical matter. When I synchronize my watch with the time I observe on the clock tower, what I am seeing is not the reading of the clock as it is, but as it was when the light that reaches my eye left its face. But how long does it take? To figure that out, I just need the distance it travels and its speed. But to find out its speed, I need to know how far it travels in a given amount of time. To determine the time requires knowing the speed which requires knowing the time. How do we get out of this circle? Poincaré has the answer—define your way out.

When we put in place the definition that accords with our intuition and ordinary usage, we are now in a place to figure out the new physical equations that govern matter and its interactions. They look very much like those that Lorentz developed, except that we now have a correction for time, and they have some incredibly odd results.

Suppose you have been appointed by the president to run the National Bureau of Standards and find yourself late for a meeting. Tying the official meter stick to the roof rack of your car, you put the official atomic clock on the passenger seat and go speeding past me. As you pass, I make quick checks of the length of your stick and the time of your clock. According to my observations, your meter stick will be the usual width, but it will have shrunk to less than a meter in length and your clock would be running slow, each tick taking more than a second. To you, however, the measurements would be spot on. It is not that the stick wrongly appeared to be less than a meter to me or that the clock wrongly appeared slow to me—these are "really" questions that Einstein has followed Mach in eliminating. Length and time are no longer open to any "really" claims; they are facts that depend on your reference frame. We can only speak of lengths or durations from someone's perspective and those lengths do shrink and durations do stretch when measured from moving frames of reference, that is, from other perspectives.

These changes are small, often too small to measure as long as the speeds are small compared with that of light, but they are there and they change with speed. The faster you go, the more I see your meter stick shrink and the slower I see your clock tick. As you get to the speed of light, the distance shrinks to zero and the time stands still. Something moving at the speed of light moves from one point to another in literally no time and all of spatial distances cease to exist.

If that wasn't weird enough, we also need to refigure how we add speeds. If I am walking up the up escalator at two miles per hour while it moves itself at one mile per hour, I am not moving at three miles per hour as Newton would figure, but slightly less. There is a correction factor that decreases the sum. The faster the speed, the more this correction factor shrinks the added speeds so that if I were moving up the escalator at the speed of light, the speed that escalator itself was moving wouldn't matter, I would still be moving at the speed of light. If I were running at the speed of light up an escalator going the speed of light, I would not be moving at twice the speed of light, but at the speed of light because of the correction factor that Einstein finds in Lorentz's work. In this way, the speed of light is a limiting velocity. No matter how fast I move and how fast the surroundings are moving, the two of them added together never pass the speed of light.

This strangeness is not limited to distance, duration, and motion. Mass, too, gets altered. The faster something goes, the more massive it is when measured by someone standing still. The closer something gets to the speed of light, the more its mass grows toward infinity.

This gives us another way to understand why nothing can move faster than the speed of light. It takes force to accelerate something. The heavier something is, the more of a push that is needed to give it a boost. If an object could move at the speed of light, its mass would be infinite. But to make an infinitely heavy object move any faster would take an infinite force, and we cannot have the infinite amount of energy needed to generate such a force. So, no object can move faster than light. This isn't an engineering problem. It's not that we haven't figured out how to do it. It cannot be done according to the basic rules governing the universe.

WITH THIS 1905 PAPER, the relativistic revolution had begun; but none of the intellectual ammunition Einstein fired had come from Jewish suppliers. It's not a foolish question to ask whether there was any Jewish influence on the theory, since the great theories of gravitation that preceded it, those of Newton and Descartes—both important and legitimate scientific theories of space, time, and motion—are indeed examples of "religious science." But in Einstein's case, there is nothing in the content of the theory of relativity that derives from, makes reference to, or was influenced by anything Hebraic. Sure, it was influenced by Einstein's cosmic religion, since it was Einstein's basic stance toward the universe that led him to think as he did. But the Nazi claim was that the theory of relativity is "Jewish science," not "Cosmic science." And when we understand "Jewish science" in this way, the Nazis were wrong. The content of the theory of relativity is not Jewish.

Of course, that is not the only way to look at Einstein's work in developing the theory of relativity.

CHAPTER THREE

Why Did a Jew Formulate the Theory of Relativity?

So, why was Einstein the one who proposed the theory of relativity? Geniuses like Lorentz and Poincaré came within a step or two of doing it. They weren't sitting behind a civil servant's desk all day, working on physics only when they could squeeze it in. These were more mature minds participating full-time in the discussions of their day, meeting with other great physicists, and reading all the big journals. Why was it this particular patent clerk without an academic appointment who was able to do it?

Could the difference be his Jewish background? Maybe it provided him with a propensity for thinking in a way that enabled the shifts needed to take that last big step. Just as a ballerina's grace requires years of stretching and strengthening starting when very young, could Einstein's upbringing in a Jewish community have provided him with the flexibility and strength of mind that was needed to jettison the luminiferous aether and absolute time?

Sadly, any answer would be pure speculation.[1] We can't read Einstein's mind much less determine his subconscious influences. But we can ask another question: were the Nazi sympathizers who championed Aryan physics correct that there is a Jewish version of the scientific method with Einstein's work as an example?

The Nazi argument has three basic premises:

1. There is a typically Jewish style of thinking.
2. This style of thinking influenced the content of the theory of relativity.
3. This influence is malicious, tainting the theory.

We saw in the last chapter that the second premise turned out to be false. The content of Einstein's work was in no way influenced by Torah, Talmud, or anything Judaic at all. The third premise is bigoted nonsense. But what about the first one? Is there a Jewish style of thinking, an approach that may not be limited to Jews or found in the work of all or even most Jewish thinkers, but which is typical of a certain type of Judaic inquiry? If so, can we find it in Einstein's science? Again, Einstein himself thinks so.

> Jews are a group of people unto themselves. You can see their Jewishness in their appearance and notice their Jewish heritage in their intellectual work, perceive a profound connection between their nature and the numerous interpretations they give to that which they think and feel in the same way.[2]

We have seen that the content of both Descartes's and Newton's theories of space were pregnant with their theologies. Can we say the same for their methods? Did Descartes think like a Catholic? Did Newton do research in a Protestant fashion? Are Einstein's advances methodologically Jewish science? The answer to all three is yes.

DESCARTES'S VORTEX THEORY OF GRAVITATION was a brilliant attempt to marry the Copernican sun-centered picture of the universe to a literal interpretation of the decree by Pope Paul V that the Earth does not move. But Descartes contended that this was no act of creativity on his part. To the contrary, he argues that it must be true, that it could not be otherwise.

Both Descartes and his brother, Pierre, were sent as children to the newly opened Collège Henri IV at La Flèche, a Jesuit institution created by the king himself. Well-funded and built on the grounds of a royal estate, it was one of the finest schools in Europe.

Pierre went as soon as he was old enough, but young René had to wait because of his tuberculosis. He received his early education at home from his maternal grandmother, who gave him access to his grandfather's library once he could read. Reading became an obsession to Descartes, whose love of letters and his physical frailty led his father to say that, like the books, young Descartes was "good only for being bound in leather."

But as soon as he was healthy enough, Descartes was sent away and quickly became a student of the highest distinction. Curious and intelligent to an extreme, he was allowed to study even topics deemed off-limits to his peers, such as alchemy and magic. Despite this great breadth and depth of study, however, he left school disenchanted, for among the works of the greatest and most penetrating minds in history there was nothing that was beyond doubt, nothing incontestable, nothing known with certainty.

Nothing, that is, with the singular exception of Euclidean geometry.

> I took especially great pleasure in mathematics because of the certainty and evidence of its arguments. But I did not yet notice its true usefulness and, thinking that it seemed useful only to the mechanical arts, I was astonished that, because its foundations were so solid and firm, no one had built anything more noble upon them.[3]

Building that more noble edifice with "the geometer's method" became his life's task.

"The geometer," of course, was Euclid. His masterpiece, *The Elements,* is second only to the Bible in number of editions published. Yet, we know very little of the man himself. He is believed to have been educated at Plato's Academy before moving to Alexandria, Egypt, where he was among the first mathematicians at that great center of learning.

The Elements is historic not only for what it says but also how it says it. Euclid took geometric results, some going back to ancient

Egypt, and organized them to guarantee their undeniable certainty.[4] He begins with basic indubitable first truths, sentences whose denial seems not even possible. These come in three kinds: definitions, axioms, and postulates. Definitions mark out what is meant by the geometric terminology, for example, "a point is that which has no part" or "a straight line is a line which lies evenly with the points on itself." Axioms, or common notions, are mathematical propositions that are used, but are not explicitly geometric, like "things equal to the same thing are equal to each other." Euclid's five postulates are explicitly geometric claims about what can be drawn, for example, that one could draw a straight line between any two points or a circle of any radius around any given point.

From these basic, intuitive first truths, Euclid derives over a hundred theorems about shapes in a plane. These are incredibly complex, sometimes counterintuitive, results proven with absolute certainty from nothing more than the self-evident definitions, axioms, and postulates.

The royal road to these theorems is deduction. Deductive arguments have conclusions no broader than their supporting premises. Consider the classic syllogism:

All men are mortal.
Socrates is a man.
Therefore, Socrates is mortal.

The conclusion attributes mortality to one man, Socrates, but the premises give it to all men. The information in the conclusion is contained implicitly in the premises. Because the content of the conclusion is narrower than and contained within the scope of the premises, it gives us a guarantee that the conclusion must be true if the premises are.

It is this guarantee that Descartes so deeply admired in Euclid and sought to reproduce in the rest of our beliefs. Showing that all of our knowledge about geometry could be developed strictly and absolutely from a small group of first principles allowed the justification of the entire enterprise to rest upon the acceptability of that small group, each member of which seemed independently obvious.

In 1637, Descartes wrote a small manuscript, *Discourse on the Method for Rightly Conducting One's Reason and for Seeking Truth in the Sciences,* designed to extend Euclid's approach to all of human knowledge. It is here that Descartes's famous dictum "I think, therefore I am" functions as the first undeniable truth, as an axiom from which to derive all other beliefs.

It acquires this status because it seems to Descartes the only proposition that we could not be wrong about. Consider the possibility that your mind is controlled by an evil demon who can feed in thoughts at will. He could make you believe anything, no matter how false, bizarre, or absurd. Descartes is not saying this is true or even likely, but if it is merely possible, then any given thought could potentially be inserted into your mind and therefore you cannot have complete confidence in anything you believe. Since Descartes is looking for the sort of absolute certainty we get from geometry, any doubt—no matter how small or stupid—is enough to derail the entire project.

So, it is therefore possible that any thought you have could be false. Any belief, that is, except for one: your belief in your own existence. The evil demon could not fool you about this because there still has to be a "you" to be fooled. To deny your own existence it is to think—denying, after all, is a form of thought—and all thinking requires a thinker. "I think, therefore I am" is a self-justifying proposition and can, therefore, function as an axiom, a first foundational truth, from which to derive the rest of our understanding.

First, he uses it to argue that an all-perfect God must exist. Descartes contends that he has the idea of an all-perfect God. Now, while the evil demon could put false ideas in his head and he could be mistaken about the truth of those ideas, he does know with absolute certainty that he is having the idea, so he cannot be wrong that he thinks about a perfect God. But such an idea is perfect and he is not. A perfect God would know everything and therefore would never doubt; but since he doubts, he must be imperfect. Yet, he has this perfect idea about God, an idea more perfect than he is. Since a less perfect thing cannot give rise to a more perfect thing, Descartes's mind cannot be the source of this idea, it must come from something else, something at least as perfect as the idea itself. But since it is absolutely perfect, the only thing that could have caused it is a perfect God himself. Hence, from his own existence as a thinking thing that has the idea of God, Descartes argues, it must be true with 100 percent certainty that this all-perfect God exists.

If God is perfect, it would be contrary to his nature to be a deceiver. As the source of his ideas, God would be the opposite of an evil demon. Based on the non-deception of God, then, we have reason to accept all beliefs that are "clear and distinct" before our minds, that is, propositions like Euclid's basic claims that are simply so obvious to us that we cannot help but believe them because they come from God. These clear and distinct ideas thereby form the indubitable foundation on which all other beliefs are to rest, especially scientific ones.

In *Principles of Philosophy*, Descartes uses deductive logic to derive his claims about space and his vortex theory of gravitation. The truth of his hybrid of Copernicus and the papal dictate follows with absolute certainty, he contends, from necessary first principles just as firmly as we know that the internal angles of a triangle are equal to two right angles. This Euclidean method is in important ways simpatico with a Catholic style of thought in that it proceeds in a top-down fashion initiated by first truths that are infallible and give rise to all other beliefs that find themselves mediated through a formal structure.

Christianity, from the beginning, certainly displayed both deep Jewish influences as well as undeniable traces of the occupying Roman Empire. Even though it arose as a contrast to the decadence of the wealthy Romans with their orgies and vomitoria, the Christian community organized itself like the Empire it opposed, adopting a stratified arrangement of bishops and archbishops mirroring the hierarchical structure of the Empire—small communities with local governors, overseen by regional leaders reporting to Rome.

In Paul's epistle to Titus, dated in the first century AD, we see the terms "elder" and "bishop" used interchangeably:

5. I left you behind in Crete for this reason, so you could put in order what remained to be done, and should appoint elders in every town, as I directed you:
6. Someone who is blameless, married only once, whose children are believers, not accused of debauchery, and not rebellious.
7. For a bishop, as God's steward, must be blameless; he must not be arrogant or quick-tempered or addicted to wine or violent or greedy for gain.[5]

This fits with the Pharisaic Jewish roots of the community in which teachers/rabbis are learned scholars but not given special powers or abilities to connect with the Divine.

But by the beginning of the second century, roles within the Church structure began to appear and a more formal position of bishop had been established. In Ignatius of Antioch's "Letter to the Magnesians," elders, deacons, and bishops are differentiated.

> Be zealous to do all things in harmony with God, with the bishop presiding in the place of God and the presbyters in the place of the Council of the Apostles, and the deacons, who are most dear to me, entrusted the service of Jesus Christ, who was from eternity with the Father and was made manifest at the end of time.[6]

"Presbyters" here is a translation of the Greek *presbuteros,* which also means elder. This is the earliest reference we have to the "monarchical episcopacy," or doctrine of one bishop presiding. This leads to the division of the land into dioceses, with bishops having spiritual jurisdiction over geographical regions by the end of the second century. This brought the structure of Church in line with that of the Empire.

In the fourth century, this relationship between church and state would move from mere structural resemblance to something much deeper. Emperor Constantine's vision of Christ's cross transformed Christianity from an abused tradition of resistance to an institution of ultimate power when it became the official creed of the Empire.

When Christianity became the state religion, it acquired the force of law to enforce its orthodoxy. The only trouble is that there wasn't actually an orthodoxy to enforce. In its first few centuries, even the most basic claims about Christianity differed widely among sects. So Constantine organized the Council of Nicea, bringing together bishops from around the world to unify Christian theology.[7]

One central concern rested on clarifying the theological place of Logos, the Word, or Jesus, the Son. Alexandrian priest Arius championed the view that Jesus should not be considered identical to God himself but rather was a creation of God. To say that Jesus is God is to suggest that two distinct Gods exist, a direct challenge to the first commandment, "I am the Lord, thy God, who brought you out of the land of Egypt, out of the house of bondage, thou shall have no other gods before Me" (Exodus 20:2-3). Arius was not denying the divinity of Christ but asserting that God and Christ were not of the same substance. A competing faction led by the Bishop of Alexandria condemned Arius, splitting the community, giving rise to riots and social unrest.

Constantine wanted the council to resolve the issue once and for all. Its findings were made manifest in the Nicene creed in which Bishops rejected the independence of Logos from God, declaring that the Son was one substance with the Father. It was as a result of Constantine's political power that Church doctrine was unified and thus the Trinity became theologically sanctioned.

While the Nicene creed clarified theological conversation among Christians, complex concerns remained. One that dogged the early Church was the place of priests, who were simultaneously the sinful flesh of humanity and also instruments of the Divine. Would the sinfulness of a priest compromise the holy sacraments he administered? The Donatists were a group of Christians who argued that such sacraments are invalid, that, in line with Paul's words in the epistle to Titus, it must be a Church of saints not sinners. But this threatened to undermine the Church's role as material mediator for the Divine. The opposition was argued by Augustine, whose reasoning ultimately won the day and helped shape the basis of Catholic theology.

Augustine was born in 354 in North Africa to a pagan father and Christian mother. In his book, *Confessions*, the first modern autobiography, Augustine recounts a misdeed of his youth that continued to haunt him throughout his life.

> In a garden nearby to our vineyard there was a pear tree, loaded with fruit that was desirable neither in appearance nor in taste. Late one night—to which hour, according to our pestilential custom, we had kept up our street games—a group of very bad youngsters set out to shake down and rob this tree. We took great loads of fruit from it, not for our own eating, but rather to throw it to the pigs; even if we did eat a little of it, we did this to do what pleased us for the reason that it was forbidden.[8]

Theft, of course, was not the only sin Augustine confessed to in the book as shown by his famed prayer, "Grant me chastity and continence, but not yet." What would lead to such unnecessary and wanton immorality?

He found an answer in Manichaeism, a combination of Zoroastrianism and Christianity in which the world unfolds as an interaction between the two axiomatic principles: good and evil. The good is the spiritual, but finds itself trapped within the bodily, the material, the evil. The goal is to preserve the goodness, liberating it from material constraints of corporeal life. Residues of Manichaeism can still be found in the character of Satan, who gets transformed in Christianity from an Old Testament figure who is a sidekick, sounding board, and enforcer for God into a competing deity who champions evil.

But Augustine was uncomfortable with the Manichean elevation of evil to a status of near divinity. He did not deny the existence of evil but thought there remained the need to reconcile good and evil with a single God. He remained philosophically troubled until he found a group of theologians locating an intellectual foundation for Christianity in the writings of Plato. It was here that Augustine saw the route out of his conundrum.

Plato, six centuries before Augustine, argues in his masterwork, *The Republic*, that most humans are like prisoners, kept their whole lives in a cave, looking only at a wall on which there are shadows.[9] Having seen nothing but shadows their entire lives, they come to wrongly believe that they are real. But one prisoner unchains himself, turning around to face the fire causing the shadows. After being blinded temporarily by the light, his eyes adjust, allowing him to see the objects casting the shadows, and he comes to realize that the world of the senses is not reality. He walks beyond the fire out of the cave and up into the sunlight. When his eyes again adjust to the brightness, he is able to study the sun and understand the true source of all light and life in the world. With this new understanding, he descends back into the cave to bring this knowledge back to those who remain trapped, chained in their false lives. Those in the cave hear his story and not only fail to believe, but, enraged by the new beliefs, kill the man.

For Plato, the world outside the cave is the world of the mind, the realm of ideas, filled with eternal, unchanging Forms, the true constituents of reality. The material world, the world of transient things that come to be and fade away, is but an imperfect representation of the true world of ideas that may be perceived only by the eye of the properly trained mind. In the allegory, the sun represents the Form of the Good, and goodness is the source of all truth and being. The man who freed himself and tried to bring knowledge of the true realm beyond the material was Plato's beloved teacher, Socrates, who was himself condemned to death by the citizens of Athens for corrupting the youth by bringing them wisdom.

For the Christian Neoplatonists, this allegory was easily reinterpreted. The sun is God and the one who emerges from the cave to later return with Truth only to be killed by nonbelievers was Christ. The bifurcation of the world into a perfect, unchanging realm in which one may find God, on the one hand, and a corrupted realm in which all is temporary and corporeal, on the other, fits naturally into the theology of Christianity. For Augustine, it provided a way to explain the evil of the world without needing another competing divine force.

It also dovetailed with original sin, which has infected every human since Eve and Adam's fateful choice in the Garden of Eden.

Original sin was not merely a physical calamity for Augustine but also a disease of the soul. Overcoming such an inherently sinful nature required God's grace to be delivered into the human heart through the Holy Spirit.

But this salvation had to occur through the Church, and the Church is comprised of humans. Augustine found himself facing the Donatists' challenge: how could sinful priests lead the way to salvation? His response was Platonic. When we speak of "the Church," we do not speak of the material manifestation—the laity, priests, and bishops that happen to make it up at this place at this time. The authority behind the sacraments comes not from the flesh of the man who is the priest but from the Church itself of which the priest is merely an imperfect material representative. The actual Church is something higher, something non-material. It is a Platonic Form, the perfect, eternal, unchanging essence of the connection between God and the world deriving its nature from the Good, from God. Sacraments are holy due to God's authority, not that of the flawed human administering them. The power of the Church transcends human affairs.

Augustine's theology ensured the security of the Church from rogue priests and the undermining ways of sin. Augustine believed that God was responsible for everything, and thus, planned everything. But he also believed that, through Church-led prayer and leading a good life, one could not only show that one was predestined to heaven, one could actually predestine oneself to heaven. That God gives grace to ward against one's sinful nature as derived from original sin exemplifies that participating in God's grace opens one up to God's love and the destiny such love brings.

But what it also does—and this is one place where Augustine's understanding of Christian theology became the bedrock for understanding both one's relationship to the Catholic Church and to Christ—is to place the Church in a special category of being, one that delivers truths incapable of coming from the sinful flesh of mortals. This philosophically reinforces the notion that the Church is the intercessor to God, that the leaders of the Church are God's human representatives, mediators on behalf of God.

So the management structure, like that of the state hierarchy, placed a supreme leader at the top of the Church who would rule from Rome with absolute authority. The supremacy of the Bishop of Rome over all other bishops throughout the Christian world was justified scripturally as an inheritance from Peter whom the other apostles deferred to, as stressed in Matthew's gospel report that Jesus gave Peter such authority. Over time, the Bishop of Rome was identified as Father, or pope.

By arranging the power of the institution hierarchically, truth flows downhill. In the Catholic hierarchy, the pope alone is the human conduit of divine revelation about matters of doctrinal faith. It was not until the First Vatican Council of 1870 that this doctrine of the infallibility of the pope when he spoke "ex cathedra," literally from his chair, was itself made dogmatic. The concept of a rigid corporate structure for the determination and dissemination of human knowledge concerning issues of faith had been in place long before that.

At the top of this structure is the pope, one might say, as CEO (chief epistemological officer) placing handpicked theologians in key administrative and advisory roles. From the upper reaches of the administration, truth works its way down through the writings and orations of those in the structure, based on what the person in the station above has said. Catholics begin with undeniable first truths and from there other truths are arranged like a telephone tree, branching down to the multitudes from a succession of levels of management.

Notice how Descartes, as a devout Catholic and a philosopher concerned with the nature of human knowledge, finds an approach for scientific reasoning that is perfectly cognate with this hierarchical arrangement. Catholic theology begins with large-scale, general, absolute truths bestowed on men infallibly from God, and all other derivative truths emerge with absolute certainty from a rigid hierarchical theological structure. In Descartes's writings, we find a scientific method mirroring this, starting again with a general absolute truth provided by the Divine that gives rise to the absolute certainty of all further derived truths through a rigid and hierarchical logical arrangement. The discovery of both sacred and secular truths proceeds along similar paths, truth flowing downhill from undeniable first truths mediated by formal structures.

Descartes's distaste for uncertainty and his resulting love for geometry may not have been religious in origin, but the eponymous method of his *Discourse* with deductive chains from indubitable suppositions underwritten by the non-deception of a perfect God is "Catholic-style" science, if not in a direct cause-and-effect sense, at least by methodological analogy.

IN THE SAME WAY, WE CAN SEE TRACES of Protestantism in Newton's writings, in both the content of his theory as well as in his scientific methodology. His intensive studies of Descartes left Newton with mixed feelings. Newton was in awe at the way Descartes had begun to mathematicize the world. Yet, he loathed the complete mechanization of the world that occurred as a result of this mathematicization. Descartes sought to understand the natural world in terms of natural causes alone. To him, the universe was a machine, and mechanical laws were to be sufficient to explain its workings. The removal of God from day-to-day operations of Creation was repugnant to Newton. The difference between their views is evident both in the theories they set out and in the logics used to justify them.

It is no accident that with the exception of a few like Galileo and Descartes, most of the great names of scientific revolution were Protestant: Antonie van Leeuwenhoek, Tycho Brahe, and Christiaan Huygens in Holland; Isaac Newton, William Gilbert, and Robert Boyle in Britain. Advances in science were of great interest to Protestants, especially at that time of great conflict between those branches of Christianity. Each new discovery both furthered human scientific knowledge and forced the discipline away from the Aristotelian views long espoused by the Catholic Church. As went Aristotle, so went the pope's credibility.

The Protestants were attracted to both the results of the new science and the means of determining them. This scientific method found its first major advocate in Francis Bacon, a lawyer and politician whose writings on education and knowledge would become the touchstone of British thought concerning the investigation of nature.

In his work *Novum Organon*, Bacon contends that Aristotle's deductive demonstrations, so beloved by Catholic thinkers, were the wrong sort of reasoning for discovering truths about the universe.[10] Science, he argued, begins with observation, with an individual interacting with the world, and then using an inductive, rather than deductive, approach to building up generalizations that outrun the data.

Induction is the logic that takes you beyond the premises to a broader conclusion. Think of polls. The pollster goes out and interviews a small subset of the population and from her results predicts the outcome of an election. If the distribution of interviewees roughly mirrors the population as a whole in terms of age, race, sex, income, and party affiliation, then the pollster can say things about the entire country, not just the few folks she spoke to. That amplification of belief is the hallmark of induction. It takes you from a finite set of data to conclusions about a larger, possibly infinite, set of circumstances. This logical leap, for Bacon, is the heart of science, it is what propels our knowledge upward toward the general laws of nature that govern the universe.

Newton built upon the Baconian view and in the middle of his *Principia* turns away from physics to discuss methodology. His "Rules of Reasoning in Philosophy" are four principles that aim explicitly at the core of Descartes's Catholic approach.[11] The first two are versions of Ockham's razor in which simplicity is a mark of truth.[12] Newton's first formulation asserts that scientific explanations should include only as many causes as necessary. His second urges us to consider similar phenomena, like the falling of apples toward the Earth and the orbiting of the moon around it, to be explained according to the same cause. These rules show why the epicycle-ridden Ptolemaic system of the ancients and the Catholics must be abandoned. Simplicity is a form of elegance. God the Creator must be maximally perfect, elegance is a perfection, and this dual aspect of God's perfection and accessibility is what the Catholics miss with their version of Aristotelianism with its absent God.

The remaining two rules contend that science is based on induction and does not move deductively from divinely supported metaphysical first truths downward to the laws of physics. Newton explicitly calls out Descartes's methodology in the *Discourse* and the absolute certainty Descartes demands:

> In experimental philosophy we are to look upon propositions inferred by general induction from phenomena as accurately or very nearly true, notwithstanding any contrary hypotheses that may be imagined, till such time as other phenomena occur, by which they may either be made more accurate, or liable to exception.[13]

The idea of science giving us results that are "accurately or very nearly true" undermines the goal and process that Descartes sets out as the basis for scientific enquiry by beginning with phenomena, with something as base as mere human observations.

Just as Descartes's process was full of the Catholic top-down approach to knowledge, so too Newton's mirrors the Protestant objections to Catholicism. Where Catholicism is inherently hierarchical with access to the word and mind of God strictly controlled, Protestantism stresses a relationship with the Divine unmediated by an organizational institution. The Latin Catholic Mass of the time made the service opaque to those in the pews, and absolution from sin requires the intervention of the institution, its priest, and rituals. But for Protestants, all are able to read the Bible, and salvation by direct interaction with God is open to everyone. The Protestant Reformation sought to undermine the monopoly on access to the Divine, opening the path to all of humanity equally in a bottom-up fashion.

The fall of Rome in the fifth century CE is seen as the emergence of the Middle Ages in part because kings began to answer to the pope. This medieval version of the divine right of kings ended a period in which earthly might established one's authority, a recipe for the chaos of constant war. When the Church stepped up as a guarantor of order, it created stability that allowed for human flourishing, at the same time solidifying the Church's power and allowing most significant political matters to rest on papal approval.

Proximity between church and state, and the need to finance both, naturally brought human corruption into the workings of the Church. The papacy found its way to the Medici family, prominent businessmen known for their patronage of the arts.[14] From 1513 to 1521 the papacy was held by Pope Leo X (Giovanni de' Medici) whose social life was as, if not more, important than religious matters. Leo was not even a trained priest. He liked to party, but he also liked to build—and both took lots of money. Like a prodigal son, Leo hemorrhaged cash. As the coffers emptied, Leo needed support for the Church, various wars, and construction projects like Saint Peter's Basilica.

To cover the costs, Leo took to the selling of dispensations, papal decrees that assured a positive outcome in terms of a deceased's place of final spiritual rest. Purchasing an indulgence from one's local priest guaranteed prayers on one's behalf, releasing poor deceased Uncle Charlie from punishment for his past sins. The selling of indulgences was so successful in raising new capital for the Church that Leo took the enterprise to previously unheard of heights—he mass-marketed them. That anyone could buy an indulgence from their corner priest did not sit well with many. Such cheapening of absolution and blatant misuse of power was appalling to much of the faithful. But it was merely symbolic of the greater corruption within the Church.

It was against this backdrop that a German monk named Martin Luther hammered ninety-five theses to the door of his church in Wittenberg in 1517 to protest the Church's use of indulgences.[15] This was not in itself radical. It was a common practice at the time for thinkers to publicly post problems for intellectual debate. Luther had no intention of offering an alternative movement to the Church with this act. Indeed, he explicitly states that indulgences are of themselves a legitimate avenue for the Church:

60. We do not speak rashly in saying that the treasure of the church are the keys of the church, and are bestowed by the merits of Christ;
61. For it is clear that the power of the pope suffices, by itself, for the remission of penalties and of reserved cases.

62. The true treasure of the Holy Gospel is the glory and the grace of God.
63. It is right to regard this treasure as most odious, for it makes the first to be last.
64. On the other hand, the treasure of indulgences is most acceptable, for it makes the last to be the first.[16]

The pope is recognized as having the power to absolve by indulgence and as long as it is a process of concern, it is a fine thing, but it is corrupted when it becomes "a favorite means for money-getting."[17] And it is on this point that the conversation began to turn: should the Church occupy the place that had been set for it? For the corruption of the Church at the highest levels had provided fertile ground for a revolution over both power and ideas within the Christian world.

As Luther began to associate his personal struggles with those of Christ, he began to separate his thought more and more from the accepted theology. His call for greater oversight of the sources of the Church's revenue transformed into a doctrinal cause. Luther's new theological understandings planted a seed that would radically challenge the structure of the Church and its role as intermediary to God. Contrary to Augustine's Neoplatonist account in which the Church is an immaterial Form, Luther and his followers viewed the Church as inextricable from the corruption, flesh, and sin of its officials.

He came to oppose the intellectual foundation of the Church's authority, that it and it alone was the conduit for God's truth and God's grace. He opposed orthodoxy in contending that God's justice and righteousness are given to all those open to God. It is not the Church or works it deems worthy that lead one to salvation, but faith itself.

> Therefore it is clear that, as the soul needs only the Word of God for its life and righteousness, so it is justified by faith alone and not any works; for if it could be justified by anything else, it would not need the Word, and consequently it would not need faith.[18]

Faith is sufficient to open one to God's grace, a theological position no longer requiring a middleman, challenging the authority and role of the Church. The importance of the institution is radically undermined. Divine truth does not roll downhill but hangs from the trees to be picked by anyone. You start by observing the Word of God and rise up to an understanding of God's truth.

This free access to God is mirrored in the Baconian-Newtonian approach to science, which offers free access to the laws of nature. Like Descartes, there is the belief in a God's-eye point of view that is available to humans who reason about it in the proper fashion. But unlike Descartes, who takes the mind and ideas from the Divine to mediate our understanding of the universe, Newton allows all of us, with the most mundane of our human faculties—seeing what is in front of our faces—to be the first stepping stone toward understanding how the Divine will governs Creation. Just as Protestant theology democratized revealed metaphysical truths allowing all equal access to the Word of God, so too the Newtonian approach to science democratizes discovered physical truths, allowing all equal access to the works of God. Science under Newton sheds exactly those aspects that are seen as consistent with Descartes's Catholicism and replaces them with versions that are structurally similar to Protestantism. In both cases, everyone has access to the starting point with his own eyes and from there one works up to absolute knowledge. In both cases, truth is universally accessible and flows uphill.

Was this relation between his theological beliefs and his approach to the scientific method intentional? Newton was certainly antipathetic toward Catholicism. Newton lived during the English Civil War when the monarchial forces clashed with the parliamentarians. Catholic hierarchical theology and political hierarchical control by the king coincided with each other and were opposed by advocates of both more diffuse theological and political philosophies of power that Protestants preferred. The connection was neither accidental, nor unnoticed.

James II worked to Catholicize the universities, most notably forcing a papist president upon Magdalene College at Oxford. When he demanded Cambridge convey the degree of masters of arts upon the Benedictine monk Father Alban Francis, there was little doubt that this was the opening salvo upon that institution's anti-Catholic

Anglicanism, something written into the very mission of the university. Newton was among the most vocal leaders of the faculty revolt against Father Francis.

Was this theological stance operative in shaping his thoughts about scientific methodology? Maybe, maybe not. It would be odd if in those troubled times in which everything was politicized along religious lines, it was not considered at all. But again the claim here is not one of cause and effect, merely one of methodological analogy. Are there ways in which Newton's style of scientific thought can be reasonably thought of as "Protestant" in flavor? The answer is clearly yes.

WHEN WE MOVE FORWARD TO THE TWENTIETH CENTURY, do we see what we saw in the seventeenth? Descartes's and Newton's theories of gravitation reflect their faiths in both the form and content. While the substance of the theory of relativity was unaffected by Einstein's Jewish background, what about the methodology, the sort of reasoning, that led him to it?

While there is certainly no direct link between Einstein's work and the rabbinic tradition, there is an interesting resemblance between their approaches to problems. We can find formal patterns that resemble Talmudic reasoning in Einstein's special theory of relativity, as well as in his general theory of relativity and his work on the photoelectric effect. Again, the claim is not that there was an influence here but rather that we can see interesting parallels in the structure of thought.

It was not Einstein who coined the term "relativity theory." Max Planck called it that and the name stuck. Einstein preferred "invariant theory," which was already a term in common usage among mathematicians.

The theory of invariants began in England with Arthur Cayley and James Joseph Sylvester.[19] Both would end up in the pantheon of mathematical gods, but when they met neither had a job as a mathematician; they were studying law. Cayley refused the academic life because at the time British universities were still church-based institutions, and one had to take holy orders to become a member of the faculty. Generally pro forma, Cayley declined because he was devout, earnest, and did not sense a true calling to the clergy. Sylvester was refused because he was Jewish.

They worked on mathematics in their spare time. Fiery, self-aggrandizing Sylvester hunted theoretical big game, the open problems of historical significance to acquire a reputation so impressive he would have to be offered a position. The modest Cayley kept a lower profile while churning out result after result.

While working on problems of algebra, Cayley noted that, when a specific family of equations was transformed in a particular way, certain properties of the solutions also changed in a predictable fashion while other properties did not change at all. Fascinated, Sylvester and Cayley developed a sophisticated theory of the properties that do and do not change. Sylvester, who referred to himself "without immodesty" as the "Mathematical Adam" for having introduced more names into the mathematical lexicon than any other human in history, termed the aspects that changed "covariants" and those that did not "invariants."[20]

If we ask whether the head of George Washington is to the right or left of Abraham Lincoln on Mount Rushmore, the answer depends on whether you are in front of or behind the faces. "To the right of" is a covariant property with respect to position; it changes in well-behaved and predicable ways when you move from one place to another. But if we say that Thomas Jefferson's head is between Washington and Lincoln, this fact is invariant with respect to position because it doesn't matter where you stand, "between-ness" does not change. We can express this sort of geometrical property algebraically, and, when equations are transformed to express the views from different vantage points, the properties of those equations can be separated into those that do not change at all and those that do change. Then we can clearly explain why the ones that vary do so in their particular fashion.

Eventually, their work gained them such recognition that Cayley was named the Sadlerian Professor of Mathematics at Cambridge, while Sylvester found prestige on the other side of the Atlantic, first at the University of Virginia (until he had to flee back to England after assaulting a student with his cane so fiercely he thought he had killed him) and then as a member of the original faculty at the Johns Hopkins University where he demanded that his $5,000 annual salary be paid to him in gold, not trusting the American currency.

Felix Klein was the second greatest mathematical mind of his generation after Poincaré,[21] living in the shadow of his contemporary. This rivalry ultimately led to Klein's emotional breakdown, but Poincaré remained unaware of it and considered Herr Professor Klein a valued acquaintance. One of Klein's great insights came as a young mathematician when he realized that the work of Cayley could be appropriated for great use in geometry, which found itself in a state of chaos at the end of the nineteenth century. No longer the noble monolith that so inspired Descartes with its patina of incontestable certainty, the introduction of non-Euclidean geometry by Nikolai Lobachevski had broken an intellectual dam.

Most of Euclid's postulates seemed self-evident. But if postulates one through four were fingers, number five was the sore thumb. It looked different. It stuck out. It seemed like a theorem, something that could be derived and not assumed. Fewer assumptions, what mathematicians call "elegance," is a virtue. For centuries attempt after attempt was made to derive the fifth postulate from the other four, to show it did not need to be assumed. All failed.

Lobachevski created a new set of axioms with a different fifth postulate. It turned out to be a strange, but perfectly possible, mathematical world. Once mathematicians realized they were no longer bound by Euclid's axioms, they created new and different geometric systems based on all kinds of different fundamental postulates. Systems proliferated at a frightening rate. Geometers were playing a mathematical version of Dungeons and Dragons, creating new and stranger realms to mentally inhabit. It became unclear, even to geometers, what geometry was anymore.

Klein realized that Cayley and Sylvester's concepts of invariance and covariance could impose order upon the chaotic world of nineteenth-century geometry. The full set of invariants of each geometric system could be used as its unique signature. By ordering these sets, seeing which sets of invariants were subsets of which other sets of invariants, the relations between the different geometries were exposed and capable of being ordered. Geometry, under what Klein called his Erlangen Program, became clean and structured.

One of Klein's star students was Adolf Hurwitz, who taught at the University at Königsberg, a small city in the Baltic, which was the hometown of Immanuel Kant. There he taught a pair of young mathematicians whom he met regularly under an apple tree near the university from which they took long walks discussing issues across the entire span of mathematics.

David Hilbert was one of these students and he would independently derive the field equations for the general theory of relativity at the same time Einstein did.[22] The other was Hermann Minkowski, a shy boy, the son of an Eastern European Jewish merchant who relocated to Königsberg when his business had gone bankrupt leaving the family with nothing. What traumatized young Hermann most was the loss of the family's library. It affected him to such a degree that he went on to memorize Goethe's voluminous works so that he would never again be without books.[23]

Minkowski's interests included both pure mathematics and mathematical physics, topics he lectured on at Switzerland's prestigious ETH. It was here, occasionally in his classroom, that Minkowski encountered a young physics student named Albert Einstein.

The relationship between them was not pleasant. Minkowski's shyness caused him to speak quietly in a halting and stammering fashion, pedagogical flaws that led Einstein, who needed little reason, to skip many of Minkowski's lectures. Minkowski thought Einstein arrogant, impudent, and insufficiently serious, referring to him as a "lazy dog . . . who never bothered about mathematics at all."[24] Einstein did not think any better of his professor Minkowski at the time, although he was later to list him among his "excellent teachers" at the ETH.[25]

Minkowski was one of the early adherents of Einstein's theory, of which he famously said, "I really would not have thought Einstein capable of that."[26] In his 1908 paper "Space and Time," Minkowski sketches out what is known as the "geometric interpretation" of the special theory of relativity. In this paper it is clear that Minkowski realized better than Einstein what the theory meant once you translated it into a geometric framework in which covariances and invariances become clear.[27] It turns out that Einstein gives us a four-dimensional account of the universe that unites space and time into a single unified space-time.

The notion of a fourth dimension is less strange than it may seem at first glance. A dimension is simply the number of numbers needed to uniquely locate something. A house needs only a single number, say 112, to be located along Mercer Street. A street is thus a one-dimensional space. To find a building in Manhattan, you need two numbers—one for the street and one for the avenue. We now have a two-dimensional space. But, of course, you would arrive at the building only to look up and still wonder where your meeting is. You need a third number, the one you push in the elevator to take you up to the proper floor. This is a three-dimensional space because you could find the location with no fewer than three numbers.

But if the interviews are being held throughout the day, you need a fourth number—the time—to differentiate your interview from those before and after yours. This is why we need to think of the universe as a four-dimensional space. To completely and uniquely locate every event, we need no fewer than four numbers, three for space and one for time.

Minkowski paints the relativistic world as one of four-dimensional cones. Every place at a time has two infinitely long cones emanating from it, one opening toward the past and one opening toward the future. The surface of this cone is defined by the speed of light. Drop a pebble on a still pond and circular ripples are created, getting larger and larger over time. The same thing happens if the flash on your camera goes off. The ball of light from the flash becomes a larger and larger sphere as time passes. In four-dimensional space, we can think of these ever-increasing circles making a widening cone when stretched over time.

Since no signal can move faster than the speed of light, the points on the surface of the cone are the limit of places at a time that you could contact or be contacted by someone. Only observers at the points on the surface or inside of the forward-opening cone can know about the event at the point of your cone, because all others would require a signal faster than light to bring them information about it and, since no signal can travel faster than light, those points are unreachable.

Similarly, all of the points on or inside of the past-facing cone are the set of all the events in the universe that an observer sitting at the point of the cone could know about at that place and at that time. These are the only parts of the past that are accessible to you then and there. All of the space-time universe outside of your past-facing cone is unknowable to you at that place at that time because an observer outside of it would require a signal faster than light to send you any knowledge of what just happened to her. Later on at different points in space-time her experience will eventually fall within your past-facing cone, and then you could learn about what happened at that place in space-time, but until you reach those other space-time points later on the information sent from her event will be unable to reach you.

In this world of cones, some quantities are covariant and others are invariant. They turn out to be very different from those of Newton, for whom measurable quantities, like length and distance, are invariant; that is, they cannot be changed. According to Newton, space is "fixed and immovable" and time "flows equably without relation to anything external." They are absolute quantities that do not depend upon perspective.

But according to the theory of relativity, this is not so. Lorentz showed that length contracts when measured from a moving reference frame. Einstein extended this, elevating Lorentz's provisional concept as a replacement for Newton's absolute notion. Duration and distance become predictable functions of perspective. Einstein then did the same for mass, taking these three basic notions and converting them from absolute quantities, invariant facts of the world, to covariant, perspective-dependent facts that change predictably with a change in one's frame of reference.

But Minkowski showed Einstein that there was something that did not change. Not everything is relative in the theory of relativity.

The theory's primary invariant is a four-dimensional measure that combines length and duration, the "space-time interval," which is not something that we can directly observe or measure. We experience the world as three spatial dimensions flowing through a one-dimensional time, but from this experience, the four-dimensional space-time interval is easily computed. And when we do, everyone calculates the same exact number no matter how fast they are moving toward or away from us in a straight line. Each of us will see different lengths and durations, but the four-dimensional combination of the two will be the same for everyone. This led Minkowski to say that with the theory of relativity, "space by itself, and time by itself, are doomed to fade away into mere shadows, and only a kind of union of the two will preserve an independent reality."[28]

At first Einstein waved off Minkowski's work. He thought it was a mere act of mathematical erudition, making his theory intellectually pretty but not saying anything new. But then his eyes opened, and Einstein realized that his professor still had something to teach him. This great paper was the class he never attended. Now he understood the structure of the universe more clearly.

And, though neither would realize it, Minkowski also showed him why his theory is really "Jewish" or, at least, "Jewish style."

WHAT WE NOW CALL "JUDAISM" emerged from Judean cultures in the first three centuries of the common era. At the beginning of this period, when the Romans governed Palestine before 70 CE and the Second Temple stood, economic and political power among the Judeans rested primarily with the priests, the Sadducees. For the Judeans, the temple in Jerusalem was the geographical and spiritual center of their religious world. Priests performed animal sacrifices and conducted services there. Judeans would make pilgrimages there for holidays or festivals, such as during Passover. Tens of thousands of Jews would ascend Zion to celebrate a liberation holiday while being watched by the occupying Romans. It's no coincidence that Romans brought in thousands of extra troops for such gatherings.

As keepers of the Holy Temple at a time of occupation, the Sadducees not only controlled the temple wealth but also needed to maintain good relations and to work closely with the Roman officials to mitigate persecution. The Romans did not select the High Priest, for example, but they did have to approve of the choice. Thus the Sadducees came to be as much a political party as a priestly class.[29] To many of the common Judeans who lived under Roman rule, the Sadducees came to be seen as selling out the rest of the tribe to protect their privileged way of life.

As Roman oppression increased, making life less tolerable, other "parties" came to the fore. Among these were the Pharisees who represented the artisans. They rejected the "go along to get along" stance of the Sadducees, explicitly opposing Roman rule through civil disobedience (at least when it was unlikely to bring down the wrath of the authorities), while a splinter group—the Zealots—preferred a more militant approach.

This split manifested itself in religious terms as well. The Sadducees took what we would now call a literalist approach to biblical texts as befits those occupying a privileged class who carries out rituals.[30] Pharisees, in contrast, had an interpretive, oral approach in which scholars taught at community-based learning centers or synagogues.[31]

These divisions tore at the Judean community. The destruction of the Second Temple would alter the cultural/political/religious landscape and gave rise to Judaism as we know it. Without the temple, the Sadducees no longer had the basis for their power.

> The loss of the Jerusalem Temple also meant that the Jewish religion had to transform itself from a Temple-based, sacrificial cult to a culture rooted in domestic and local practices; celebration of the annual cycle of agriculturally based holidays; and the transfer of the purity laws to the home and to the house of study . . . The synagogue liturgy developed beyond its Second Temple origins and became the basis for all Jewish liturgy in subsequent centuries.[32]

With the loss of the temple, the Pharisees became the dominant party, and their oral, interpretive approach became the way of Judaism.

But with decentralization from the loss of the temple came worry. Would distant groups lose the wisdom that had come from the scholars? This was a real concern. To help alleviate this fear, the rabbis wrote it down, creating the Talmud.

The Talmud, the interpretive companion to the Torah, is made up of two parts. The first is the Mishnah, which explains how to exercise many of the blessings or commands of the Torah. Long passed on orally, the Mishnah was codified in the second century CE under the leadership of Rabbi Judah Ha-Nasi, although the groundwork was first laid by Rabbi Akiva, who systematically organized all of the Jewish legal commands, *halakhah,* into meaningful bits.[33]

Like the Constitution, rabbis found the codified Mishnah vague when they tried to apply it to real-life cases. This inspired rabbinic discussions and commentaries, which developed into a process of searching the Mishnah and Torah for meaning over a four-hundred-year period. This commentary is gathered into the second part of the Talmud, the Gemara, which functions like Supreme Court opinions interpreting the rules. In the Torah and Mishnah, God is absolute but can only be understood through interpreting and performing God's words. This can only happen in real-life situations.

In order to both sustain tradition and disrupt the propensity of tradition to suffocate the new on behalf of the old, an interpretative method known as Derash (literally, to seek), or Midrash, developed to simultaneously continue tradition while refreshing it. When the biblical and mishnaic words, commands, and traditions are midrashically mined in seeking to understand what God wants from us, we are exposed to God in the sense in which an archaeologist is exposed to an ancient culture when a pottery shard is discovered, showing the continuity between civilizations.

Rabbis are not taking liberties with the word of God in midrashically interpreting it but rather disclosing novel aspects of it. God's truth is bigger than our contextual experience can expose, more multifaceted and intricate than our interpretations, but through them and the actions they require, new facets of it are revealed. The word of the text and the performance of the commands are observed and inspired through Midrash, but the Absolute is never fully observed.

Traces are accessed, glimpsed through the text and through performing the commands for the good of another.

Consider the seemingly straightforward command, "Thou shall not steal." In the real world, this simple notion can quickly become convoluted. Thus we get Talmud's tractate (chapter) Bava Metzia, or "Middle Gate":

> Which found objects are his, and which is he obliged to announce? These found objects are his: [If] he found scattered fruit, scattered coins, small sheaves in the public domain, or round cakes of pressed figs, loaves of a baker, strings of fish, or pieces of meat, or fleeces of wool that have been brought from (lit., "their") country, or stalks of flax, or strips of purple wool, these are his. [These are] the words of Rabbi Meir.[34]

and the related interpretive passage in the Gemara:

> All these articles are the finder's to keep, because they appear identical to others of the same kind and bear no distinguishing marks by means of which their owners can identify them. We therefore assume that their owners have given up hope of getting them back and have abandoned them. The above is the opinion of Rabbi Meir.[35]

The guiding principle here is finders keepers *if* the original owner has given up hope of finding it. So, if the object has no distinguishing marks by which the owner can identify it, it follows the owner can have no hope of recovering it.

So, is it stealing to keep a hundred dollar bill I find on the ground? It depends. If I find it in a mall parking lot on December 23, when thousands of cars are around and no one in sight is looking for it? No. In a library? Maybe. In the hallway of a friend's house? Definitely. Same commandment, same action, but circumstances matter in determining whether the action follows the word of God.

This contextualization is part of rabbinic training. Students approach the Talmud with a study partner, a *chevrutah,* from the Hebrew word for friend, *chaver.* Without a teacher imparting knowledge through authoritative lectures, the partners' discussion drives them through the issues in the text as they debate and react to one another. This interchange of ideas, thoughts, and philosophies creates a learning environment that trains the participants to see every issue from many sides. The goal is not merely to arrive at a single truth but also to create multiple insights that inspire better behavior in a wide variety of circumstances. They discover that the Truth is bigger than any single interpretation and, while it can only be lived through a single life's contexts, full sense can only be made of it through understanding multiple standpoints.

God's Truth is so big that it is mysterious and must be explored with a companion to ensure against getting trapped in your own limited understanding. The truth is never fully revealed to one of us, because we can only survey the horizon from our own perspective. Each partner—through discussion, disagreement, and agreement—shares in the Absolute, getting a glimpse of the Truth from an oblique angle relying on your *chevrutah* to reveal more to you than you could see from your own vantage point.

There is an analogy to draw between Talmudic interpretation and the theory of relativity. The heart of the Talmudic view is that there is an absolute truth, but this truth is not directly and completely available to us. We can only see it through our experience, which is limited to a context. In our search for deeper meaning, we must try to understand how that limited view of the truth fits together with seemingly contrasting views of the truth from other different perspectives and contexts. It turns out that exactly the same style of thinking occurs in the relativity theory and in some of Einstein's other research in the period.

Let's work again through some of the reasoning in Einstein's seminal paper, "On the Electrodynamics of Moving Bodies,"[36] the article that introduced the theory of special relativity. Recall that Einstein begins with the thought experiment in which we have a circuit with no battery but that includes a coil of wire and a magnet that fits inside it. First, we hold the magnet still and move the coil back and forth around the magnet. This gives rise to an electrical force that creates an induction current in the circuit. Next, the coil is held still while the magnet is moved back and forth inside of it. This creates a dynamic magnetic field that gives rise to an equivalent current.

Under the old theory, Maxwell's electrodynamics—named for Scottish physicist James Clerk Maxwell—the identical effect is given two completely different physical explanations depending on which is really moving, the coil or the magnet. If we adopt the Newtonian style of thought in which there is a single absolute truth that humans can access through observation and induction, then these two distinct descriptions tell us that the world is really different depending on which one is really moving, because the state of the aether is different in the two cases. These are absolute answers to the "really" questions. The theory gives us the one true account of the state of the aether that serves as the privileged viewpoint, as God's frame of reference, from which one determines the underlying facts of reality. Newton thinks he can read the mind of God and fully understand whatever he reads.

But that's not what Einstein says. Because the two cases give you the exact same observable result and only differ in terms of the point of view, he sees them as different glimpses of the same reality. There is no problem with Maxwell's theory or the myriad of conflicting experiments trying to determine whether the aether is dragged along with the moving Earth like Descartes's space or whether the Earth moves through it like a butterfly net through the air as we would have with Newton's absolute space. The problem is that the scientists before him have all been doing *goyische*-style science. They think there is an absolute state of the aether.

The problem isn't in the science, it is in the interpretation. If we see the different accounts as contextualized glimpses of the same bigger underlying truth, problem solved. Don't give up Maxwell's theory, just the picture it paints of the world with a privileged frame of reference, a God's-eye view that humans can occupy. By adopting a "Jewish-style" approach in which we only experience the world from a limited context and, therefore, have an incomplete view of the larger absolute truth, each perspective giving us different, but no less true, results, suddenly we have no need for an aether at all. When we start with the relativity principle, that all frames of reference are equivalent, then there is no need to try to find the golden one that explains everything else and so no need for the aether.

The Lorentz transformations are based on Maxwell's theory, which holds the speed of light to be constant regardless of the relative motion of the source. So, keeping Maxwell's laws forces us to replace Newton's theory of motion in which a speed could never be constant in different frames of reference. In this new theory, distances, durations, and masses become covariant quantities. There are no "really" questions to be asked about them. Truths we thought we knew about an object's mass or size remain truths, but now they are truths of a more limited sort tied to frames of reference. As in the Talmud, there are both overarching absolute truths and also contextualized truths that are no less true, but whose domain is limited to a particular context, to a reference frame. Einstein was an intellectual mohel who, by circumscribing the metaphysically absolute status of space and time, made physics "Jewish style."

Minkowski performed the theory's bar mitzvah, putting it in its mature form, providing the Gemara to Einstein's Mishnah. Minkowski's understanding of the theory in terms of a four-dimensional space-time continuum allows us to understand that there are different truths about the length of a moving object from different points of view. We can also see that these interdependent truths are to be understood in terms of a larger absolute truth only available from a point of view that is inaccessible to humans trapped in the world of perception with three-dimensional space and one-dimensional time. There are absolute truths, but these express themselves differently depending on the circumstances.

When you, the head of the National Bureau of Standards, speed by me with a meter stick tied to the roof of your car and an atomic clock on the seat next to you, the length and duration we observe are different for you than they are for me. But the four-dimensional space-time interval that we both calculate from the different observational data is exactly the same. It's just that since we do not live four-dimensionally, we never partake of the experience of the universal, absolute truth even if we can calculate it from our experiences. We live through our three-dimensional lives in which we experience this absolute truth in a contextualized fashion. In this way, the theory of special relativity is methodologically "Jewish-style" science in a fashion similar to the way that Descartes's physics is methodologically "Catholic-style" and Newton's methodologically "Protestant-style."

This "Jewish style" of thinking is not unique in Einstein's work on special relativity but can also be seen in another of his papers from the miracle year of 1905, his paper "On a Heuristic Viewpoint Concerning the Production and Transformation of Light"[37] in which he offers an explanation for the photoelectric effect.

Understanding the nature of light was an ancient problem. When we realized that light moved, there seemed to be only two mutually exclusive options. One, championed by Newton, was that light is made up of particles, which he termed "corpuscles." Particles occupy a place and move around by themselves. Newton's contemporary Christiaan Huygens advanced a rival theory in which light is a wave. In Huygens's view, waves are disturbances in a medium and do not have an existence independent of that medium—no water, no water waves. Waves are not localized but are spread out through the medium. The light of distant stars travels to us through the vast vacuum of space. This makes the particle view seem more plausible since there is no medium to convey a disturbance, to do the waving.[38]

But in the first half of the nineteenth century, experimental data suggested that light was a wave. The wavelike behavior was unmistakable in circumstances like shining light through thin slits. The resulting light and dark bands were indicative of waves adding and subtracting in ways that particles could not. (Recall our discussion of wave function from chapter 1.) Maxwell's theory showed that optical phenomena could be explained as an electromagnetic wave propagating through the luminiferous aether and could predict the observed behaviors.

Einstein, of course, was in the process of doing away with the aether in his theory of relativity, which left him with the need for a new account of the nature of light. He took advantage of a problem that the wave theory could not account for—the photoelectric effect.

When an ultraviolet light shines on metal, electrons are kicked out. That is not strange in and of itself. The outermost electrons of metal atoms are loosely bound—that's why metal wire is a good conductor of electricity—and if light waves hit the metal, the surface would vibrate (think of a struck tuning fork brought near an unstruck one) and the vibration would liberate the most precariously held electrons on the metal's surface.

Making the light brighter should affect the electrons emitted. If schoolchildren are holding a blanket with ping-pong balls on top and start to wave the blanket, the balls will bounce up. If they make bigger waves, the balls will go higher, they will leave the blanket moving faster. If light is a wave, then brighter light is a wave with a larger amplitude and that should make the electrons come off the surface of the metal moving faster.

But they don't. More electrons come off, but they are emitted at the same speed. The classical account with light as waves through the aether couldn't make sense of it.

Einstein pulled in a result from Planck who made sense of another problem—what he called "blackbody radiation"—by hypothetically postulating that light behaved *as if* it came in discrete bundles of constant energy.[39] Light, Planck argued, should be treated in this one case like eggs and not butter. You can buy any amount of butter you want. No matter how small the pat, you can cut it in half. But eggs, you have to buy in ones. You can't buy half an egg. Planck could account for the spectrum of emitted blackbody radiation, if we think of light as if it were eggs not butter, as if it were made of particles not waves.

This was the same "as if" that Lorentz made use of with his notion of the contraction of lengths. And just as Einstein thought in a new way about Lorentz's work to give rise to the theory of relativity, he did the same with Planck's thoughts concerning light. It's not "as if" light comes in little quantized packets, termed photons, it actually does. Barbra Streisand's voice may be like butter, but light isn't—Einstein showed us that it's like eggs.

The photoelectric effect makes sense, because we no longer have kids waving a blanket with ping-pong balls on top; now we have a pitcher throwing baseballs at the surface. Electrons are kicked out whenever a pitch hits one. Since light travels at a constant speed, making the light brighter would not mean the pitches are thrown harder, but rather that more pitches are thrown. More pitches means more electrons kicked out, and since the speed of all pitches is the same, the energy of the emitted electrons should be the same. Problem solved.

But in solving this problem, we are left to ask whether light *really is* a wave or a particle. We still have all the earlier examples of wave-like behavior, but now there are Planck and Einstein's particle-like cases. Which is it really, particle or wave?

A "Christian-style" question asking for an absolute answer. A question Einstein responds to in a "Jewish-style" way: it is a wave from some perspectives and a particle from others. Wave-particle duality allows us to adopt different points of view based on our frame of reference. This does not mean there is no absolute truth about the nature of light; it merely means that the way in which we understand the behavior is dependent on the context in which we ask the question.

This is the foundation of the theory of quantum mechanics. Indeed, it is for this work that Einstein received his Nobel Prize. And it is a second example of "Jewish-style" science in the methodological sense appearing in Einstein's work.

Yet another instance is found in Einstein's general theory of relativity.[40] The move to the general theory came because Einstein had two problems with the original. First, the relativity principle only held good for reference frames that moved at constant speeds with respect to each other. The real theory ought to be open to any reference frame whatsoever—even accelerating ones. Second, while the original theory accounted for almost all of physics at the time—mechanics, optics, electrodynamics—there was one force missing: gravitation. How could it be worked in?

The trick for both involves what he termed "the principle of equivalence." Suppose you are in an elevator at the top floor of the Sears Tower standing on a bathroom scale. You push down on the top of the scale, the floor pushes back, the springs inside squish, and it reads your weight. Suddenly, the unthinkable happens. The cable snaps and in your last moments hurdling down the shaft, you look down at the scale. What does it read? Zero. Why? Because you are on top pushing down, but the floor is no longer pushing back. The scale and the floor are falling at the same rate as you are, so there's nothing to squish the springs.

Now suppose you are in a rocket on your bathroom scale, far out in deep space nowhere close to any planet or star. Because there is no measurable gravitational force, the scale reads zero. Then the engines roar to life, and the rocket accelerates upward. The floor now pushes up on the bottom of the scale and the springs start to contract. If the acceleration is just right, the scale will read your weight on Earth.

So, if you wake up to find yourself in a small metal room with nothing but a bathroom scale, could you tell whether you were in an elevator on Earth or a rocket far out in space? If the scale reads your weight, you could be at rest in a gravitational field or accelerating upward in a rocket in zero gravity. If the scale reads zero, you could be at rest with no gravity or accelerating downward where there is gravity. But which one is it really?

Really? Another "Christian-style" question that demands a "Jewish-style" answer. Einstein argues that acceleration and gravitation are the same thing, just viewed from different points of view. Each reference frame will divide up the effects of acceleration and gravitation differently, but the combination is invariant. There is an absolute truth, but we never see all of it. We are only left to glimpse it from our own context.

This equivalence led Einstein to posit a curvature of space and, when the concepts from the special theory of relativity are applied to this curvature, we get his general theory of relativity—yet more methodologically "Jewish-style" science.

Is this result trivial? Every set of physical equations has a set of transformation equations with properties that are covariant and others that are invariant. Does this mean every theory is "Jewish style"? Newton's laws are invariant under what physicists call the

Galilean transformations, so is Newton's work the scientific version of Jews for Jesus?

No, because for Newton there is absolute motion and relative motion, and the two are not to be confused. Newton argues quite clearly that, while relative motion is "common, base, and merely apparent," "absolute, true, and mathematical" motion is determinable by its "causes, effects, and apparent differences": that is, motion relative to absolute space itself. From God's vantage point, the thing is moving or it isn't. Even though there is a degree of relativity in the symmetry of the equations when transformed from certain viewpoints into others, truth is not uncovered with each different point of view contributing a piece, the whole truth being larger than can be perceived from any perspective. Rather, the truth is there, you just have to know which one is the golden reference frame from which to perceive it.

What is "Jewish style" about Einstein's approach in devising the theory of relativity is his commitment to the existence of an absolute truth that can only be glimpsed through limited perspectives. This notion of a larger truth, expressing itself in the world through context, is formally similar to Talmudic discourses in a way we do not see in the work of Descartes and Newton. But is it to be found in the work of all other Jewish scientists?

Let's take the other most famous Jewish scientist who worked at the same time as Einstein, Sigmund Freud. Biographically, his intellectual development is remarkably similar to Einstein's. He was also a secular Jew who considered himself to be part of the Jewish community. "My parents were Jews, and I have remained a Jew myself."[41]

His family fled persecution, seeking refuge in Austria, where he distinguished himself as a student and selected medicine as his profession. Here, like Einstein, he came to find himself labeled by anti-Semitism in the schools.

> When, in 1873, I first joined the University, I experienced some appreciable disappointments. I found I was expected to feel myself inferior and

> an alien because I was a Jew. I refused absolutely to do the first of these things. I have never been able to see why I should feel ashamed of my descent or, as people were beginning to say, of my race. I put up, without much regret, with my non-acceptance into the community; for it seemed to me that in spite of this exclusion an active fellow-worker could not fail to find some nook or cranny in the framework of humanity. These first impressions at the University, however, had one consequence which was afterward to prove important; for at an early age I was made familiar with the fate of being in the Opposition and being put under the ban of the "compact majority." The foundations were thus laid for a certain degree of independence of judgment.[42]

As with Einstein, the experience of anti-Semitism in education led Freud to embrace being the outsider.

> My language is German. My culture, my attainments are German. I considered myself German intellectually, until I noticed the growth of anti-Semitic prejudice in Germany and German Austria. Since that time, I prefer to call myself a Jew.[43]

It did not manifest itself in a resentment of learning or knowledge, but a sense that insight was not to be found with the small-minded authorities who claimed it.

Anti-Semitism made Freud an intellectual revolutionary, and he clearly cites it as a functional factor in his ability to do what he did.

> Finally, with all reserve, the question may be raised whether the personality of the present writer as a Jew who has never sought to disguise the fact that he is a Jew may not have had a share in provoking the antipathy of his environment to psycho-analysis. An argument of this kind is not often uttered aloud. But we have unfortunately grown so suspicious that we cannot avoid thinking that this factor may not have been without its effect. Nor is it perhaps entirely a matter of chance that the first advocate of psycho-analysis was a Jew. To profess belief in this new theory called for a certain degree of readiness to accept a position of solitary opposition—a position with which no one is more familiar than a Jew.[44]

Freud asserts as clearly as Einstein both that there are indelible traces of his background on his way of thought and that the German-speaking world would oppose his work on the grounds of his background.

But it was not just the cultural aspects of Judaism that were at work.

> My early familiarity with the Bible story (at a time almost before I had learnt the art of reading) had, as I recognized later, an enduring effect upon the direction of my interest.[45]

Unlike Einstein, religious themes occupy a significant place in Freud's writings, but unlike Descartes and Newton, whose theological orientation led them to create theoretical systems consistent with their extra-natural commitments, Freud uses his theory to try to explain away such stances.

> Religion is an attempt to get control over the sensory world, in which we are placed, by means of the wish-world, which we have developed inside us as a result of biological and psychological necessities. But it cannot achieve its end. Its doctrines carry with them the stamp of the times in which they originated, the ignorant childhood days of the human race. Its consolations deserve no trust. It is a dramatization, projected into cosmic order, of sentiments, fears and longings that develop from the relationship of child to parents. If one attempts to assign religion its place in man's evolution, it seems not so much to be a lasting acquisition, as a parallel to the neurosis which the civilized individual must pass through on his way from childhood to maturity.[46]

Freud's hostility toward religious belief is clear and, while religion is certainly a prominent subject matter, in no way can we call the content of his view "Jewish science" in the sense discussed in the last chapter.

But what of our metaphorical, methodological sense of "Jewish-style" reasoning? Freud acknowledges that his place as Jewish outsider was a factor in his thinking. But do we also find the sort of

Talmudic-style invariance-covariance that we saw in Einstein? It is promising at first glance since psychoanalysis requires two people, the patient and the analyst, who are in dialogue and clearly occupy distinct frames of reference while trying to find a single practical response to a real-world problem that will depend on the patient's biographical context.

But this view of psychoanalysis as "Jewish-style science" is betrayed by Freud's own understanding of it. Unlike the Talmudic study partners or the magnet and coil in Einstein's thought-experiment, there is not an equivalence between the perspectives of the patient and the analyst. There is not a single inaccessible truth that we can only glimpse through limited perspectives, but which we can come to understand in a larger context by understanding how it changes with the viewpoint. Rather, for Freud, there is a fact of the matter about the source of his patients' neuroses within the contents of the patients' unconscious according to Freud and it is through the psychoanalytic methods that the analyst receives a privileged view of that truth. The analyst, in a metaphorical but very real sense, is playing God—or at least pope—and that ability for a single person to occupy a position of knowledge of the absolute truth is indicative of a "Christian-style" not a "Jewish-style" methodology.

Consider his case of the patient Freud referred to as "Rat Man."[47] While on vacation with his girlfriend, he began compulsively exercising, running for hours in the heat of the day up and down mountains, and dieting in a very unhealthy way. Why was he suddenly struck with the overwhelming urge to lose weight?

His girlfriend was staying with her American cousin named Richard who went by the nickname "Dick." Rat Man saw how his girlfriend adored her cousin and became extremely jealous. Subconsciously, Freud argues, he wanted to kill Dick to get him out of the way. Of course, this desire could not be consciously formed by someone living in a society in which murder is frowned upon, so it manifested itself differently. It just happens that "Dick" is the German word for fat. So, Rat Man's subconscious desire to get rid of Dick was transformed into a desire to get rid of fat. This made him obsess in a way that seemed utterly unexplainable, but once we understand the workings of the subconscious mind, Freud contends, it becomes clear what actually happened.

This case, which is typical of Freud's work, is not methodologically "Jewish style" in the sort of way that Einstein's work is. There is a single correct answer to questions about the source of Rat Man's behavior. It is accessible from a single privileged viewpoint, that of the therapist. It's not the case that Rat Man's conscious experience gives us a part of the truth and his subconscious gives us another glimpse that must be reconciled from an unattainable human perspective. Freud sits in the privileged place from where the whole truth may be discovered.

Indeed, Freud sees the spatial position as indicative of one's position to the unseen truth.

> My patients, I reflected, must in fact "know" all the things which had hitherto only been made accessible to them through hypnosis; and assurances and encouragements on my part, assisted perhaps by the touch of my hand, would, I thought, have the power of forcing the forgotten facts and connections into consciousness. No doubt this seemed a lot more laborious process than putting them into hypnosis, but it might prove highly instructive. So I abandoned hypnotism, only retaining my practice of requiring the patient to lie upon a sofa while I sat behind him, seeing him, but not seen myself.[48]

Certain facts about the patient's life are not known to the patient but are known to the unseen seer who then can reveal the underlying mystical facts of the patient's soul. Just as Newton thought he was reading the mind of God in developing his laws of motion from empirical data, so too psychoanalysis allowed careful listening and experiments like word association to divulge the contents of a patient's subconscious. Indeed, Freud casts himself in a popelike role of being the sole receiver of the hidden truths, the absolute facts which then can be fully revealed to the suffering mortal. As such, Freud's psychoanalytic approach is much more methodologically "Christian style" than it is "Jewish style." So, not all Jewish scientists do "Jewish-style" science.[49]

But some, like Einstein, did. Another Jewish scientist in whose work you will find this "Jewish style" of reasoning is Emile Durkheim, one of the founding fathers of sociology. He came from an observant family. His father, grandfather, and great-grandfather were all rabbis in France, and he himself was sent to rabbinic school, but rejected it. His secular interests led him to become a teacher of teachers. He worked very hard to modernize pedagogical practice in France, trying to make it much more scientific both in content and approach. In thinking about the ways in which people are affected by institutions such as schools, Durkheim was able to introduce sociology to French schools and became the first sociologist in France, teaching first at Bordeaux and later at the Sorbonne in Paris.

His central contentious claim is that there are social facts, facts about human culture that are not reducible to biological facts or psychological facts but exist alongside of them. His first major work that illustrated this was *Suicide,* which examined the reasons for the sudden and remarkable increase in the number of deaths by suicide that were occurring in the last decades of the nineteenth century.[50] On the one hand, nothing could be more personal a decision than to take one's own life. There is no doubt that in considering any particular suicide, we have to ask about the person's mental and physical health. The reasons why this person made the choice he did has to do with biography, biology, and state of mind.

On the other hand, Durkheim asks, is it not incredibly odd that the rate of suicide in a country is directly correlated with the percentage of Protestants within the country's borders? It turned out that the lowest rates of suicide were to be found in European countries that had the highest concentrations of Catholics. As the percentage of Protestants rose, so too did the suicide rate. Similarly, it was well known that Protestants were by in large wealthier than Catholics and more likely to be better educated and hold positions of power and authority in the private sector. While the decision to take one's own life is surely a personal one, there are clearly, Durkheim contended, larger social factors at play that make some people more likely to act in certain ways than others. In addition to reasons from the person's health and psychological state, for a complete explanation, we also need to account for the social facts.

Durkheim works out what he means by social facts in his book *The Rules of Sociological Method,* which lays out the object and process of sociology.[51] Here he argues that social facts are defined by two elements. First, they are objective things that give rise to certain thoughts, actions, or emotions in the individuals of that society. When you go to a movie, people laugh, cringe, or jump out of their seats at the same time by the same images. That you found a certain line of dialogue funny is clearly something internal to you, but it is an interesting fact that most audience members also react in similar ways to the same thing at the same time.

Second, social facts come with coercive force. People who do not act or react in the ways that others do are subject to social sanction. It may be imprisonment, it may be a disapproving glance, it may be detention, or simply not getting dates, but when you act in a way that opposes a social fact, there is a price to be paid. After a while, we often lose sight of the fact that we are acting in accord with a social fact as it becomes a natural part of how we think and make sense of the world. We wear clothes in public, not because we are afraid of arrest or the strange looks we might get, but because we just automatically get dressed without thinking about it before leaving the house. This sort of socialization, the internalization of social facts, Durkheim argues, is part of what we do in education—for better and worse.

This is one reason why we often fail to see these social facts for what they are. They become part of the lens through which we see the world and we look through lenses, not at them. They become so ingrained in us that we think them natural laws, not social constructions and why we as humans are naturally conservative. We get used to the way things are done, internalize it into our very being, and then even if the change would be rational and helpful, we oppose it because it is a social fact we have incorporated into ourselves and we implicitly know what happens to those who violate social facts.

This approach to sociology is methodologically "Jewish style" in the same way that Einstein's theory of relativity is. For Durkheim, while social facts are things, we never see them. If I tell you that it is a fact that I am wearing blue pants, then the sentence "I am wearing blue pants" is true of something, in this case, me and my pants. But what are social facts true of? I know how to point to myself and to my pants, but in the case of the social fact that people in this society are required to wear pants, what do I point to? What it is true of? The society. But what in the society can I point to show its truth?

Durkheim's answer is subjective, internal experience.

> When I fulfil my obligations as brother, husband, or citizen, when I execute my contracts, I perform duties which are defined, externally to myself and my acts, in laws and in custom. Even if they conform to my own sentiments and I feel their reality subjectively, such reality is still objective, for I did not create them; I merely inherited them through my education.[52]

The social facts are objective things of the world; they are the subject of the study of sociology. But to get at them, we have only the lens of the subjective personal experiences of people living in that society.

See how "Jewish style" this is. There are objective realities, a way things really are, but we never see them directly in their entirety. They are glimpsed only through the lived, contextual experience of individuals. Sociologists get at them obliquely through observations of individuals, of their thoughts, emotions, and actions, but they can only understand these realities for what they truly are by trying to think across perspectives since objective reality will be manifested differently in each context and from each person's position.

So, unlike Freud with his privileged place from which the analyst can see the real truth about a person's psyche, Durkheim argues that there is no God's-eye view for the sociologist, that we can strive to find the real, existing objective truth about the world only by understanding how it is instantiated in the very different lives of different people. Just as Einstein gives us a methodologically "Jewish-style" physics, so too does Durkheim give us an analogically "Jewish-style" method of sociology.

We therefore have an interesting sense of "Jewish science" that captures a number of works, including but not limited to Einstein's. Again, this is a metaphorical sense of the term "Jewish." There is no claim here that it is in any way related to Torah, Talmud, or anything else connected with Jewish history or customs. This is not to claim a causal effect on Einstein, Sylvester, Durkheim, or any other Jewish scientist or mathematician. Indeed, you may be just as likely to find this approach in the works of non-Jews as you would with Jewish thinkers. But when we contrast it with the absolutist, top-down, deductive approach of Descartes or the absolutist, bottom-up, inductivist approach of Bacon and Newton, the metaphor does hold. It does look a lot like the style of reasoning we see with traditional Jewish thinkers. So at least metaphorically, relativity can be thought to be "Jewish science" in one sense.

CHAPTER FOUR

Is the Theory of Relativity Political Science or Scientific Politics?

We usually take the word "Jewish" to refer to a specific historical, cultural group bound together by beliefs, traditions, and customs. We saw in the first chapter that, even in this narrow sense, the word is still multiply ambiguous. But the situation is yet more complicated, because words have not always meant what they mean to us now. Einstein's theory of relativity was called "Jewish science" in Germany between the world wars. So far we've been trying to see how we can make sense of that phrase in light of our contemporary understandings of "Jewish."

But for German speakers at the end of the nineteenth and beginning of the twentieth century, "Jewish" was a loaded term; it had other connotations suggesting a whole lot more. The Aryan physics movement emerged from a cultural context in which "Jewish" moved from implying a specific non-Aryan group to having the sense of anti-Aryan and of being antithetical to everything that is authentically German. Jews went from being an Other to *the* Other. What caused this change?

The easy answers invariably use the term "scapegoat." This is no doubt true, but why this scapegoat and not some other? Of all the minorities, why did "Jewish" become the one that had clear oppositional meaning, especially in a country in which much of the indigenous Jewish population was so assimilated that they viewed themselves as German before Jewish, if Jewish at all. The anti-Semitism in German culture was not terribly different from what you would find at the same time in, say, France or the United States. But something happened. To make sense of the development of the Aryan physics movement, we'll need to unpack three connotative aspects: Jewish as unhealthy, Jewish as disloyal, and Jewish as modern.

IT IS IMPORTANT TO UNDERSTAND that modern Germany is much younger than the other prominent nations of Europe around it. Einstein was born only eight years after Otto von Bismarck united Germany into a coherent political unit in 1871. The Germany of Einstein's formative years was a pubescent nation undergoing rapid and profound changes and experiencing the growing pains that inevitably accompany such a state of flux.

At the beginning of the nineteenth century, Napoleon had conquered the area, wresting it from the control of the Holy Roman Empire, and had set up under his "protection" a weak confederation based on the French model of government. While it never really functioned as a representative democracy, the intent was clear—to import French liberal Enlightenment values to the German-speaking world. It resulted in a split in German society: conservatives, on the one hand, who opposed all things French and who longed to return to traditional aristocratic rule, and reformers, on the other, who used the writings of thinkers like Immanuel Kant to "Prussianize" the Enlightenment and argue that there needed to be greater individual rights, freedoms, and responsibilities for all.[1]

When Napoleon's army was pushed back by the Russians, and he met his final defeat at Waterloo, the wave of anti-French and nationalistic feelings that swept over the German-speaking regions made their residents eager to destroy anything that smelled of the former French occupation. The parliament was dissolved (not that it ever actually met anyway). The liberal ideas behind it were vilified, and the area became a loosely bound collection of autonomous regions and free cities, each with its own nobility monopolizing local control, the northern states of the alliance always looking nervously at Austria, the powerhouse to the southeast. This led the northern regions to begin to cooperate a bit with each other, at least on economic and tariff issues in a way that led to increasing interconnectedness without Viennese influence. At the same time, land disputes with France and Denmark gave rise to a "rally 'round the flag" instinct, resulting in increasing pan-German sentiment.

When tensions with Denmark boiled over in 1866 regarding the disputed duchies of Schleswig and Holstein, and the Seven Weeks War was won by the full German-speaking confederation, victory caused more conflict. Austria, concerned about regional isolation, had wanted the newly liberated areas to be independent, but Prussia, the largest and most powerful of the northern states, had insisted on annexation. A war erupted and Prussia, to the surprise of nearly all, won quickly and decisively in the aptly named war under the leadership of Bismarck.

Victory over Austria galvanized the northern states, but Bismarck sensed that the tipping point was still yet to come. There was one thing that would truly bring all of Germany together—their hatred of the French. Knowing that German unification was intensely feared by France, Bismarck goaded Napoleon III into declaring war on the German states with some geopolitical saber rattling and half-true press releases that creatively misquoted the French ambassador. When a unified northern German confederacy defeated the despised French in the Franco-Prussian war and occupied Paris in 1871, Bismarck seized the moment to install Wilhelm I, the Prussian king, as the leader of a united Germany. His title was Kaiser, a germanized version of Caesar, not only implying that the position was imperial in its scope and authority but also overtly harkening back to the pre-Napoleonic period, twin sentiments that would warm the hearts of German conservatives.

Bismarck then wrote a new constitution creating the Second Reich in which the Kaiser selected a chancellor who had complete executive control over all governmental affairs domestic, foreign, and military. This chancellor, of course, was to be Bismarck himself. In addition, there was a parliament with an upper house, the Bundesrat, and a lower house, the Reichstag, that would be popularly elected.

Unification intensified the wave of nationalistic sentiment that tied military prowess to the major economic and social restructuring that was occurring. Industrialization lengthened life spans and created significant new wealth among those who had not been blessed with it previously, but centralization of workers in factories also meant mass migration out of the countryside and into the cities. Leaving farm work in the fresh air for factory work in the polluted urban environment gave rise to a public health disaster.

The backlash this spurred in the popular mind is tied to a romanticized image of nature and the broad-chested, blond-haired German men who lived in harmony with it, an image infused with Nordic mythology and rabid nationalism. This image was contrasted with the city-dwelling Jew, a weakling who for his own profit lures the fit and bold German away from the land of his fathers into dark, cavernous factories causing harm to the healthy, vibrant constitution of Germans for the purpose of enriching his own pockets.

We see this line in the works and writings of Richard Wagner. In his most famous work, the cycle of operas *The Ring of the Nibelung*, Wagner takes us to a Nordic mythological world in which the hero, Siegfried, knows no fear as he faces down gods and monsters in his quest for a magical ring. Siegfried is brave, beautiful, and in touch with nature. He can speak to the birds and lives among the trees and mountains. The antagonists are the *Nibelungen*, ugly troll-like dwarves who live below the earth—literally as subhumans—and who plot and scheme to take from the gods and giants who live on the Earth. One of these dwarves, Alberich, who made the magic ring, schemes to get it back. But knowing that he is not strong enough, literally not man enough, Alberich raises the orphaned Siegfried in his cave with the express purpose of having Siegfried kill the ring's current owner, Fafner, so that the dwarf can get the ring.

It is not a big leap to see Siegfried as the archetypal German and the *Nibelungen* as Jews.[2] Indeed, it would be like modern audiences seeing Robert DeNiro or Joe Pesci on the screen and not immediately identifying the character as Italian. Wagner is clear in his essay "Judaism in Music" that the Jew is an "unpleasant freak of nature" whose "exterior can never be thinkable as a subject for the art of re-presentment."[3] Jews are the opposite of the vital, handsome German. "We can conceive no representation of an antique or modern stage-character by a Jew, be it as hero or lover, without feeling instinctively the incongruity of such a notion."[4] Where the Nordic types are the superhuman image of health like Siegfried, Jews are the subhuman epitome of sickness like Alberich. So we should see the conniving Jews as trying to control the stronger, heroic Germans in order to secure riches they are neither deserving of nor able to gain on their own.

We see this German as healthy, Jewish as unhealthy dichotomy again clearly stated a decade later in the writings of Friedrich Nietzsche. In his book, *On the Genealogy of Morals*, Nietzsche traces our ethical notions back to preethical conceptions. Think of the words we use to morally praise those who do good deeds, we call them "gentlemen" and their acts "noble." But, of course, these are not really ethical terms, they are class indicators. The words we use for that which is good are really terms that were originally used to indicate that something was of the powerful. Similarly, the words we use to morally condemn an act are terms like "common," "base," or "low," words that denote lower class, those without power.

Nietzsche argues that this is not accidental. Originally, these terms were mere class indicators having no ethical connotation at all. When the world was ruled by a group Nietzsche calls "aristocrats," the word "good" just meant "good for me," and when someone was condemned as bad, it just meant "I'm glad I'm not you." These aristocrats were not people of conscience, they were men of action. Like Wagner's Siegfried, the aristocrats held an active notion of ethics, of living, "filled with passion through and through—'we noble ones, we good, beautiful, happy ones!'"[5] They had "small horizons," living completely in the present moment, not letting the past fester inside of them, but always seeking new and greater challenges through which to triumphantly affirm their existence in the world. "One cannot fail to see at the bottom of all these noble races the beast of prey, the splendid *blond beast* prowling about avidly in search of spoil and victory."[6] Their values were healthy for both spirit and body.

But these values were turned upside down when our current morality was born. How did this happen? Nietzsche sets it out as a general principle that morality is a function of power. Whoever is in charge of the culture gets to not only make the rules and enjoy the riches but also to define the moral vocabulary. The aristocrats started by defining themselves and what they do—satisfying their physical urges—as good, and then defined the opposite—not having your needs met—as bad.

But there was a slave revolt, a reversal of power that resulted in an inversion of values. For the aristocrats, good was active: getting, doing, being. But when the weak took control, things reversed. For the new bosses, the priests, ethics is ruled by prohibitions, by thou shalt *not*.

> Think, for example, of certain forms of diet (abstinence from meat), of fasting, of sexual continence, of flight "into the wilderness": add to these the entire antisensualistic metaphysic of the priests that makes men indolent and overrefined.[7]

The proper way to live according to these priests is not to satisfy your deepest desires but to abstain from satisfying them, to live a life of constant denial. It is restraint that is prized.

But, of course, you will still have those physical urges; that is just part of being human. The wanting and the abstinence lead to frustration and repression. Repression gives way to self-loathing and resentment toward anyone who gets anything. In this way, the aristocrats' notion of "bad" turns poisonous and transforms itself into the concept of "evil"—something that had never existed in the simple and pure approach to life of the aristocrats. Morality, as we know it, according to Nietzsche, is born of unhealthy self-loathing and removal of ourselves from our bodies and our nature, that is, alienation from all that brings health.

And who did Nietzsche think it was that caused this destruction of the physical and spiritual well-being of humanity? Who was responsible for this unhealthy inversion of values?

> All that has been done on earth against the "the noble," "the powerful," "the masters," "the rulers," fades into nothing compared with what the *Jews* have done against them; the Jews, that priestly people, who in opposing their enemies and conquerors were ultimately satisfied with nothing less than a radical revaluation of their enemies' values, that is to say, an act of the *most spiritual revenge*. For this alone was appropriate to a priestly people, the people embodying the most deeply repressed priestly vengefulness. It was Jews who, with awe-inspiring consistency, dared to invert the aristocratic value equation (good = noble = powerful = beautiful = happy = beloved of God) and to hang on to this inversion with their teeth, the teeth of the most abysmal hatred (the hatred from impotence), saying "the wretched alone are the good; the poor, impotent, lowly alone are good; the suffering, deprived, sick, ugly alone are pious, alone are blessed by God, blessedness is for them alone—and you, the powerful and noble, are on the contrary the evil, the cruel, the lustful, the insatiable, the godless to all eternity; and you shall be in all eternity the unblessed, accursed, and damned" . . . In connection with the tremendous and immeasurably fateful initiative, provided by the Jews through this most fundamental of all declarations of war, I recall the proposition I arrived at on a previous occasion [*Beyond Good and Evil*, section 195]—that with the Jews there begins *the slave revolt in morality*: that revolt which has a history of two thousand years behind it and which we no longer see because it has been victorious.[8]

It is the Jews who have transformed German culture in a way that leads the once great and proud, the noble and aristocratic, to become unhealthy. Jews are impotent themselves and they actively transform the culture in order to cripple the virile blond beast, forcing it into their unhealthy ways.[9]

This notion was presented to the German people in high art and philosophy and in popular culture. Starting in the 1890s, a series of cheap paperbacks by Karl May became hugely popular, especially with young boys.

> Their influence on young peoples' thinking should not be underestimated. They are set in a world of clear-cut black and white distinctions, and the main character is young, manly, noble in character and German. With or without the help of some noble savage the hero fights for order, humanity and victory of good over evil—and is invariably successful. Boundless patriotism, love of nature and pseudo-Christian religious feelings form the underlying themes of the stories . . . There are, incidentally, many autobiographies which document the lasting effect Karl May's books had on the thinking and attitudes of the young men who read them.[10]

Not unlike the influence of Daniel Boone or the Lone Ranger in American popular culture in the 1950s, the heroes of Karl May's stories captivated the imaginations of German youth and fed into a romanticization of a life lived in harmony with nature.

As these children got older, love of these types of stories gave rise to the *Wandervogel* and Free German Youth movements in which groups of teens would organize themselves in small cells to hike, camp, and get back to the land. These groups would reject standard social norms of dress, often embracing nudism. They would sing traditional folk songs to try to reconnect with a preindustrial German identity and seek out farmers in the countryside whom they held up as ideals to emulate.[11]

While this is somewhat reminiscent of the hippie movement of the late 1960s, there was a significant difference. Where the American counterculture had anarchic and democratic foundations, trying to live in a communal fashion, disdainful of authority, the Free Youth, on the other hand, embraced authority, each group focused around a charismatic leader who needed to constantly defend his place and who became endowed with a mystical aura.[12]

These groups originally covered the political spectrum. But as they became more widespread and their social influence became clearer, the leaders started to shift from older adolescents to adults. With this change, the movement became largely co-opted by the political right with the rhetoric from them becoming more and more overtly anti-Semitic.[13] Once again, the idea of land and health gets tied to an image of Jews as their despoilers.

We see this formalized with the rise of the Third Reich, who became the first government to enact proenvironmental legislation.[14] In 1934, they passed the *Reichsjagdgesetz*, or national hunting law, that banned among other things the use of hounds in hunting, and the *Reichstierschutzgesetz*, or the national animal protection law, that forbad cruelty to animals and animal experimentation. The next year, it was further extended in the *Reichnaturschutzgesetz*, the national law for the protection of nature, in which the natural world was to be seen as intrinsically valuable, not just of utilitarian usefulness. Nature and the health that came with living in accord with it was a German value undermined by Jews, who were identified from Wagner to Hitler as vermin and rats—carriers of disease.[15]

WHILE ONE ELEMENT OF THE CONNOTATION OF "Jewish" was alienation from the natural world and the health it brings, another prominent sense is Jew as disloyal. Jews were a peculiar minority. Identity in the age of empires came from flags. Whose borders do you live within? Whose language do you speak? To which crown are you loyal? Humans could be partitioned into sets of subjects correlated with nation-states. But Jews were part of a group that had no state. They were everywhere but had nowhere. Germans could mistrust the French. French could mistrust the British. The British could mistrust the Irish. It all made sense because borders were not exact, and there was something real, something material at stake in the conflict. Protestants could accuse Catholics of disloyalty in a way that was reasonable. They preferred the authority of the pope to the local sovereign. The pope is a person, Rome is a place, and the Catholic Church is an institution as real as any government. But what is it that the Jews obey? Who is their authority? Where and to whom is the loyalty of the Jews directed? Where do we need to invade and whom do we need to overthrow to conquer them?

Jews broke the traditional categorization that was used to construct the normal picture of the world. For this they were to be distrusted in a deep way. Jews, by being spread across the globe but not having a locus of control, represented a notion of internationalism at a time when it had previously not meaningfully existed. But this cosmopolitan, internationalist posture found adherents among the two sociopolitical movements that most threatened German conservatives: communism and liberalism.

Following Friedrich Hegel, Karl Marx argued that history is a deterministic process working itself out according to a rigorous, inviolable logic. The state of things at any time inevitably gives rise to its exact opposite. Of course, a thing and its opposite will not coexist peaceably and so there is conflict. If one of the two were stronger, it would win the battle, but since these are mirror images of each other, the two destroy each other and from the ashes arises the next more advanced state of reality that contains within it the deeper truth that was contained within, but masked by the contrast in, the previous state. The process continues until a final complete state is reached, resulting in the end of history.

For Hegel, this historical dialectic was the mind of God, the Absolute, Spirit, or *Geist* working out its own nature. Is God the knower or the thing known? History ends when God becomes unified with himself as both subject and object. All is God and the history we see is just a manifestation of the mind of God as he works out the nature of his own Being.[16]

Marx contended that reality is not a mind-state of the Divine but rather the stuff of the world. Hegel's God is a fiction. The material stuff around us is the furniture of reality. We are in a world of people and things, things made by people, and people made into things. Science studies the behavior of things, and since people are themselves also material, we need a science of society. Society has a structure and creates structures of its own, structures in which power comes from political position but also from wealth. Political power, something real but ephemeral, is inextricably connected to the material possession of land, stone, and metal. We cannot separate the political from the economic, and history is the stepwise series of struggles of class against class.[17]

Having a notion of history that privileges the concept of class allowed Marx to call for the workers of the world to unite in their common struggle. In the Marxist vision, the workers in Germany were workers first and Germans by accident. Identification no longer relied on flags but on a completely extranationalist quality. Marx was not holding one nation to be superior to any other but rather undermining the entire category of national identity in the same way the presence of the Jews threatened to. If Nietzsche killed God to let Germans be Germans, then Marx killed God to turn everyone into Jews.

But it was worse than that. The bad guys of Marx's narrative are the bourgeoisie. It was one thing to try to undermine the place of the *Bürgertum*, the traditional German gentry, the old-fashioned wealthy with long pedigrees. Everyone, even the *Bürgertum* themselves, was piling on as the new modern, powerful, technologically advanced Germany was emerging.[18] The *Bürgertum* were fuddy-duddies, and their ways had to go. But the new bourgeoisie, whose wealth, power, and place were the result and symbol of German ascendance, were flourishing because Germany itself was flourishing. They were rewarded because they were heroic in the German mold.

Marx's transformation of the new bourgeoisie into the enemies of history instead of its apex was a direct affront to the idea that Germany was moving into its rightful place as dominant world power, militarily, economically, and artistically. It was the most heinous form of treason to change the rules of the game just as the home team was about to win. And it was lost on none of their opponents that Marx himself and a number of his most prominent advocates came from Jewish families. Indeed, Hitler himself refers to him in *Mein Kampf* as "the Jew, Karl Marx."[19]

But German communists were not the only ones who were taken to illustrate the disloyalty of Jews. The same fate occurred to those associated with liberal democracy. Every country has a national mythology that sets out positive virtues, descriptions of what it is to be great and why that country embodies them in aspiration, if not actuality. But at root, these are really negative claims about the faults at the core of those who needed to be defeated to give birth to the nation. England, France, and the United States all threw off the shackles of monarchy with its ossified political structures in which unqualified heirs were given authority and wealth they neither deserved nor used wisely. Because of an accident of birth, these royals were held to be superior to those who worked harder, were smarter, and more just. This drove the thinkers who created the constitutional structures in these societies to concepts introduced during the Enlightenment, the claim that "all men are created equal" is a self-evident truth and should serve as the basis for governmental legitimacy. As Nietzsche correctly points out, a revolution becomes invisible when it triumphs completely, and so we fail to see how radical an act the sending of the Declaration of Independence was, how shocking it must have been for mere commoners to say to King George III that it is self-evident, that only a moron would deny that, they are equal to him, that the basic assertion of the power of the king, that God himself places the king above his subjects, is not only illegitimate, but that no one with any sense could accept it.

But the German revolutionaries of the nineteenth century were not opposing a monarchy; they were freeing themselves from Napoleon's liberal France. Liberty, equality, and fraternity were the values of the enemy, and so, the German national mythology evolved in contrast to it. People are not created equal. Quite the opposite, people are inherently unequal: there are heroes and there are sheep. What liberal democracy does is to give the power to the masses, to the weak and mindless. It keeps the few, the great, the strong, the healthy, from taking their rightful place. Rule by the weak will infect the entire culture with weakness. The great should rule and the society and institutions will reflect their strength and self-assuredness. In this vein, conservatives loved the writings of Arthur Schopenhauer:

> A peculiar disadvantage attaching to republics—and one that might not be looked for—is that in this form of government it must be more difficult

for men of ability to attain high position and exercise direct political influence than in the case of monarchies. For always and everywhere and under all circumstances there is a conspiracy, or instinctive alliance, against such men on the part of all the stupid, the weak, and the commonplace; they look upon such men as their natural enemies, and they are firmly held together by a common fear of them. There is always a numerous host of the stupid and the weak, and in a republican constitution it is easy for them to suppress and exclude the men of ability, so that they may not be outflanked by them. They are fifty to one; and here all have equal rights at the start.

In a monarchy, on the other hand, this natural and universal league of the stupid against those who are possessed of intellectual advantages is a one-sided affair; it exists only from below, for in a monarchy talent and intelligence receive a natural advocacy and support from above.[20]

This spirit was projected onto Wilhelm II. When his grandfather, the original Kaiser, died in 1888 and his father, Frederick III, passed three months later, Wilhelm II became the Kaiser and sacked Bismarck, whom he despised for taking all of the power away from his grandfather. He had no interest in letting the chancellor have control, consolidating it instead in the position of the Kaiser.

Wilhelm was born with a withered left arm. Like Napoleon's obsession with his height, Wilhelm felt the need to compensate for his own anatomical lack, both physically, by always having his left hand on a sword or some other instrument to make it seem usable, and psychologically, by being boisterous, aggressive, womanizing, and attention-seeking in every context. His need to be adored and his self-absorption led to his inability to listen, something that rendered him politically inept to a stunning degree. He gathered sycophants rather than insightful ministers around him and was well known for changing his mind with no warning and for seemingly no good reason.[21] Wilhelm loved technology and thought that German industry and German militarism were inextricably linked to one another and formed the path to the place which God himself had reserved for them, as the great global power. Military parades became a common part of life. The German chemical industry dominated the world market. With chests out and heads held high, conservatives finally had their Germany.

When Wilhelm led the German people into the Great War, a war that he painted as a preemptive strike, a necessary defensive struggle, the population whipped itself into a nationalistic frenzy, all but a few token internationalists swept up in the mania. So unanimous was the war furor that Wilhelm declared that he no longer recognized parties but saw only Germans, a statement that was greeted with near universal approval. He sought a victorious peace, a *Siegfriede*, whose Wagnerian overtones were obvious.

But the war did not go as planned, and Wilhelm was forced to abdicate. German cities were not occupied. Its citizens were controlled by no foreign power. The war, it seemed, could not have been lost. Yet, Wilhelm was gone and with him the structure of the Second Reich.

It was replaced with a republic, of all things, a parliamentary republic in the style of England or, even worse, France.[22] When the love of your life leaves you for another man, it is to be expected that you will not be fond of her new beloved. But if you are a lifelong francophobe and she leaves you for a Frenchman, the anger and hatred will be intense. And so nationalists felt extreme disapproval of both the form and the content of the new government when it became parliamentarized with the elevation of the Reichstag—a body that Wilhelm had referred to as an "imperial monkey house."[23] To make matters worse, the new constitution drafted at Weimar was a product of the left of center Social Democrats, the centrist German Democratic Party, and the aptly named Center Party that drew centrist Catholics into its ranks.[24]

This was the government that would negotiate peace with Germany's enemies, establishing terms the conservatives thought unfair. This was the government that would have to raise taxes to pay off the reparations it negotiated, taxes that caused serious inflation. And this was the government that put Jews in positions of power and reversed longstanding anti-Semitic laws. The socialists had become the majority party, yet ceded leadership of the coalition government to the newly formed German Democratic Party, founded by liberal German intellectuals with an eye toward "rational republicanism," eschewing the bombastic rhetoric of the far left and right.[25]

The November 18, 1918, edition of the widely read paper the *Berliner Tageblatt* held an appeal to form the party contained a number of widely recognized names printed as signatories. Among them was Albert Einstein. The party had wide appeal for the middle class who saw it as a safe harbor from extremism. It received funding from a number of sources including electronics magnate Carl Friedrich von Siemens. But significant among its funders were major Jewish industrialists. It attracted overwhelming Jewish membership and included Jewish members in positions of power. It became easily tagged by the German right as the "Jewish party." And when it was placed at the head of the ruling coalition, the entire government was seen as "Jewish."[26]

Most provocatively, Walther Rathenau was put in the post of foreign minister, the highest governmental position ever held by a Jew and one that immediately raised suspicions on the right who were already concerned by the internationalist orientation of the government. Rathenau was the son of the founder of the German General Electric, AEG, and was himself a powerful industrialist and extremely successful corporate manager. Rathenau bragged that all of Europe's business was conducted by a group of three hundred men, a group to which he belonged.[27]

But success in the boardroom was not what motivated him. Rathenau was an intellectual, author, philosopher, and member of the circle of prominent Berlin avant-garde thinkers and artists. Edvard Munch painted his portrait. He was often disdainful of nonassimilated Jews, yet was open about his background,

> In the youth of every German Jew, there comes a moment which he remembers with pain as long as he lives: when he becomes for the first time fully conscious of the fact that he has entered the world as a second-class citizen, and that no amount of ability or merit can rid him of that status.[28]

He felt himself a modern German, longed to be the quintessential modern German, yet always knew that for all of his great talents and achievements he was condemned always to be an outsider to the culture he knew, loved, and considered a defining characteristic of his own being.

Rathenau was one of the early adherents of the German Democratic Party and donated generously to it. He held various positions before his promotion to foreign minister. He met with representatives of the British government to try to modify the terms of reparations payments and was the first representative of Germany to meet with Soviet ministers, working out a small treaty between the two countries that opened formal relations, something that worried the powers of the West but that Rathenau thought he could finesse in face-to-face meetings. Six weeks later, on June 24, 1922, Rathenau was shot dead, assassinated by members of the right-wing *Freikorps* militia.

He was killed because he was a Jew and because he represented a stance toward the rest of the world that German conservatives fundamentally opposed. Jews were not to be trusted as true Germans because they were Jews; their loyalty was not to the Fatherland but to no land. Jews were internationalists and therefore to the eye of the archnationalist, inherently disloyal.

THE THIRD ASPECT OF THE TERM "Jewish" to the gentile German ear between the world wars is being "modern." In this context, modern not only stands in opposition to traditional, normal, and comfortable but also brings with it the negative connotation that is afforded the word "new" in the phrase "new math" or "new Coke": that is, it is taken as a pejorative by those who see it as moving away from what was tried and true and throwing what was taken for unmistakable truth into a false cloud of doubt.

A standard anti-Semitic stance is that, in addition to being greedy, Jews are able to manipulate others for their own gain. Whether it is Wagner's thinly veiled *Nibelungen* or Marx's claim that the secular essence of Judaism is hucksterism,[29] there is the sense that the Jew is a double-talker who brought out needed wariness. Some translate Marx's term as "bargaining" instead of "huckstering," but the sense is well known to all that the stereotypical Jew is one who will try to, and often succeed in, using words to confuse you and put you at a disadvantage. The Jewish mind is clever in a dangerous way. As Nietzsche puts it, "his spirit loves hiding places, secret paths, and back doors."[30] Jews will seductively lead you into a linguistic labyrinth and once there you are easy prey.

The huckster is not merely a liar but what philosopher Harry Frankfurt has termed a "bullshitter."[31] A liar is one who believes something and tries to get you to believe it is false. There is, Frankfurt argues, a commitment to the nature of truth in lying. To lie is to believe that there is a truth of the matter but to want someone to not believe it for reasons of personal gain.

> The liar is inescapably concerned with truth-values. In order to invent a lie at all, he must think he knows what is true. And in order to invent an effective lie, he must design his falsehood under the guidance of that truth.[32]

But bullshit, he argues, is much more pernicious. The bullshitter is not restricted in having to make his deception consistent with agreed upon truths. He does not have to engage us in generally accepted facts of the world, but in the act of his huckstering or bullshitting is creatively free of the strictures of truth altogether. In this way, the bullshitter not only speaks falsely, he rejects truth itself and the world to which it supposedly refers.

> It is impossible for someone to lie unless he thinks he knows the truth. Producing bullshit requires no such conviction. A person who lies is thereby responding to the truth, and he is to that extent respectful of it. When an honest man speaks, he says only what he believes to be true; and for the liar, it is correspondingly indispensible that he consider his statements to be false. For the bullshitter, however, all these bets are off: he is neither on the side of the true nor on the side of the false. His eye is not on the facts at all, as the eyes of the honest man and the liar are,

> except insofar as they may be pertinent to his interest in getting away with what he says. He does not care whether the things he says describe reality correctly. He just picks them out, or makes them up, to suit his purpose.[33]

The Jew was viewed by the anti-Semite as despicable for the nature of Jewish reasoning, which denies the world itself in the way Frankfurt much later describes.

> Astonishingly, truth and reality do not appear to be anything at all special or different from untruth to Jews, but are equivalent to any one of the many different . . . options available.[34]

The anti-Semites argued that this lack of commitment to a robust notion of absolute truth, along with being an element of Jewish activities in the marketplace, is to be found in the Talmudic forms of argument at the heart of rabbinic discourse. Among the anti-Semites, there is no appreciation of a richer theory of knowledge in which one has to account for perspectival elements of a larger absolute truth beyond the direct access of the limited human mind. Instead, it was contrasted with the German Romantic metaphysic, which is tied to nature and being with nature in a way that is simple and authentic. Thus, the complexity of Jewish thought was seen as treacherous over-refinement that led to alienation from that which is real and true. Indeed, such danger was, in Weimar culture, to be found everywhere.

After the Great War, a deep sense that humanity stood at a turning point in history came from many directions: rapid industrialization, political and social upheavals, new wealth and leisure, youth culture consciously and vocally breaking away from tradition, and trench, chemical, and mechanized warfare that slaughtered sixteen million. People knew that, for better and for worse, the future would not resemble the past. That sentiment forced a radical reconsideration of the foundations of all beliefs, such as what is it to be a human? How do we determine what is right and what is wrong? What is beautiful? How should we organize ourselves socially, politically, economically? Does anything have more than a transient meaning?

This state of what historian Fritz Stern labeled "cultural despair"[35] gave rise to simultaneous and interrelated changes in almost every intellectual and artistic endeavor. No longer were these undertakings exclusively outward looking but instead artists, scientists, and humanists were forced to turn their intellectual microscopes on themselves and their own work. Challenges to traditional forms, constraints, and methods proliferated. Nicholai Lobachevski's non-Euclidean geometry took what was regarded as the most solid intellectual creation and threw it into doubt. Pablo Picasso took the two-dimensional canvas, which had for centuries been used to depict the third dimension, and tried to instead represent four-dimensional realities.[36] Arthur Schoenberg composed in a way that intentionally disregarded the notion of key, long considered the fundamental building block of music. Expressionist movements in literature, drama, and film undermined standard approaches and basic structures, while Bauhaus architects made buildings look like nothing that had been seen before. Freud's psychoanalytic theory reworked what it was to be a conscious being. There was nothing that one could take for granted anymore. Anything you tried to lean on as an unmovable foundation for certainty underlying any belief or practice was being shifted or eliminated by someone.

And one could not help but see Jews at the vanguard of virtually all of these movements in fields throughout the intellectual landscape. The outsiders who had a knack for thinking in convoluted ways had invaded every aspect of modern life and scrambled it all, leaving nothing as we always thought it should be. So, conservatives rejected this modern thinking wholesale, yearning for a return to "authenticity," and easily classified the new modernism as Jewish and thereby pernicious.

> As early as the turn of the century an element began to push its way into our art which up to that time could be looked upon as entirely alien and unknown. Perhaps in previous times errors of taste happened sometimes, but the cases involved were artistic derailments to which posterity at least gave a certain historical value, rather than products of degeneration which was no longer artistic at all but rather senseless. Through

them the political collapse, which later on, of course, became better visible, began to announce its arrival in the cultural field.[37]

While Hitler only discusses artistic endeavors here—Hitler was, after all, an artist himself—in the same way, elements of the German right equally condemned as "Jewish" modern works in everything from philosophy to mathematics and, of course, physics.

The hallmarks of modernism and its "Jewish influence" are (1) the rise of self-referential work (art, for example, not being about a beautiful thing, but about what it is to be a work of art), (2) its appeal to symbols and formalism instead of the world itself (looking for nonstandard forms, like non-Euclidean geometry, in which to couch counterintuitive theorems generated from a sort of linguistic game), and most especially (3) its inaccessibility and complexity (films, plays, and novels that made no sense and music that displeased the ear that leave the viewers or listeners feeling as if they are not "in on the joke" and thereby diminished in a game of false refinement). In every matter of modern life, it seemed, there were challenges to the traditional ways of doing things and it seemed as if Jews were not only leading the charge, but benefitting from this rethinking. As such, the term "Jewish" carried with it for German-speakers in the lead up to the Second World War a connotation associated with modernism as they yearned to return to the old, storied ways.

WE WILL SEE ALL THREE of these notions—Jew as unhealthy, Jew as disloyal, and Jew as modern—rolled into the concept of "Jewish science" as it was applied to the theory of relativity, but there were more focused biases in play as well. Other branches of modern physics, most notably quantum mechanics, would also be tarred with the label "Jewish," but there was something about relativity that made it scientific public enemy number one. That something was Albert Einstein.

World war was the nightmare Einstein had feared since his childhood. He fled the militarism embedded in the culture at sixteen, renouncing his German citizenship because of it. He reluctantly agreed to come back to Berlin as an adult, persuaded by Max Planck with an offer that was too tempting to pass up, but still harboring all of the hatred for kneejerk nationalism he remembered from his youth. He saw the German conservatives not just as distasteful but as dangerous. Theirs was not idle bluster; it was leading somewhere, somewhere awful. So when Wilhelm led the nation to war, Einstein saw the carnage and destruction through an angry lens thickly coated with "I told you so." The zeal had swept up friends, people he loved and respected who were vocally supporting a militaristic adventure for its own sake. As the war proceeded, Einstein's public criticism grated on those who didn't want to hear it. When the whole enterprise turned out to be less than successful, the anger and humiliation only made Einstein's barbs sting more intensely.

It didn't help matters that after the war this gadfly would become one of the world's most recognizable figures. Sir Arthur Stanley Eddington's observations that provided the first major new evidence for the general theory of relativity made global headlines. Einstein, the antiwar traitor, the Weimar supporter, was being heralded by the world, especially the enemies who killed Germany's brave sons and were now bleeding the Fatherland dry from reparations.

The resulting hatred of Einstein by German conservatives was akin to that of American conservative's view of Jane Fonda during the war in Vietnam. As Johannes Stark put it:

> Since the end of the war the French have suppressed the German people in the most brutal manner. They have torn away piece after piece from their body, have engaged in one act of extortion after another, they have placed colored troops to watch over the Rhineland, and they have made insufferable demands on the German people through the reparations commission. And just at this very time, Herr Einstein travels to Paris to deliver lectures.[38]

Einstein was, for conservatives, the symbol of everything that was wrong with the culture. He represented the very processes that were undermining traditional values and ways. And the source of his power was this bizarre theory that any reasonable mind, they contended, would see for the nonsense it was. If that theory could be undermined, then so would Einstein, and the destruction of the symbol of change would itself be a symbol of the reascendance of the real Germany.

And so the attacks began on Einstein's theory. A pamphlet was printed called "One Hundred Authors against Einstein" in which twenty-eight writers from across the intellectual spectrum, from chemists to engineers to philosophers, contributed short essays, a couple of paragraphs to a couple of pages, explaining why they objected to Einstein's theory.[39] They cited others, some physicists of renown like Erich Kretschmann and Paul Ehrenfest, who had voiced concerns. Any objection was good enough to mention, as long as it meant another writer against Einstein.

And the attacks continued in public forums and publications across Germany. Paul Weyland, an anti-Semitic engineer and right-wing activist, and Ernst Gehrke, an experimental physicist in Berlin, organized the "Working Group for German Scientists for the Preservation of Pure Science," which sponsored free lectures throughout Berlin with the express purpose of undermining popular opinion about relativity theory. After attending one of these talks, Einstein felt he had to respond, not because there was any threat to the theory itself—indeed he considered both "unworthy of a reply from my pen"[40] and the whole syndicate a "motley group—but because it seemed to be getting out of control and he needed to set things straight once and for all.

So, with classic Einstein snark, he wrote an editorial for the *Berliner Tageblatt* on August 27, 1920, called "My Reply. On the Anti-Relativity Theoretical Co., Ltd." His argument is reminiscent of Socrates's defense at his trial in the *Apology*. Socrates contends that while he is innocent of the actual charges brought against him, the real reason he is being tried was for being an annoyance to the state in daring to oppose the powers that be.[41] Similarly, Einstein writes,

> I have good reason to believe that there are other motives behind this undertaking than the search for truth. (Were I a German nationalist, whether bearing a swastika or not, rather than a Jew of liberal international bent . . .)[42]

But, again like Socrates, after dismissing what he sees as the real charges, he undermines the claims against him, pointing out where his accusers misinterpret his theory, use discredited data, and mislead their listeners.

In doing this, Einstein not only picks on Weyland and Gehrcke but also singles out another critic, Phillip Lenard, an experimental physicist, who won the Nobel Prize in 1905 (the year Einstein first published on relativity) for his work on cathode rays.

> I admire Lenard as a master of experimental physics; however, he has yet to accomplish anything in theoretical physics, and his objections to the general theory of relativity are so superficial that I had not deemed it necessary until now to reply to them in detail.[43]

He takes Lenard's argument that relativistic results do not accord with everyday experience and points out that at the speeds we see in everyday life, the results of his relativity theory reduce to Newton's—something that would be obvious to any professional. Then—pairing it with a flawed and "utterly worthless" attempt to derive the advance of the perihelion of Mercury (one of the major pieces of evidence in favor of the general theory of relativity) from Newton's physics by school teacher Paul Gerber—Einstein rails:

> The personal attack Messrs. Gerber and Lenard have launched against me based on these circumstances has been generally regarded as unfair by real specialists in the field; I had considered it beneath my dignity to waste a word on it.[44]

Einstein had abstained from engaging before with these misguided, politically motivated critics out of concern that it would generate a circus instead of an environment for serious science. So he should have known better than to conclude with a direct challenge to them:

> Finally, I would like to note that on my initiative arrangements are being made for discussions to be held on relativity theory at the scientific con-

ference in Nauheim. Anyone willing to confront a professional forum can present his objections there.[45]

This was a mistake. Announcing an open forum to opponents and singling out Lenard would both change the tone and elevate the issue. If Einstein was trying to stop his science from becoming a cultural proxy battle, he had just failed spectacularly.

The meeting in Bad Nauheim, the first general meeting of the German Physical Society after the world war, threatened to turn into a riot. Weyland's group seized on Einstein's challenge and showed up organized and in significant numbers.[46] Police were called to surround the meeting hall. Inside, the session was run by Planck, a colleague and friend of Einstein's, but also a nationalist conservative who, like Lenard, had signed the "Manifesto to the Civilized World" defending the German invasion of Belgium. Most of all, Planck was a true Prussian gentleman who stood on restraint and decorum. So, when Weyland's group repeatedly and in a clearly organized fashion disrupted Einstein's remarks, Planck, "pale as death," sternly quieted the gallery.[47]

But Lenard was not with Weyland's group and did not share or approve of their tactics. Lenard was there to engage in serious and authentic scientific debate over an issue he thought controversial and wrong. The unruly and inappropriate antics of the antirelativity protesters, however, led other physicists to think that Lenard was their spokesman. He was thereby disdainfully dismissed by his peers. This insult, a Nobel laureate being treated as less than a real physicist, was one decisive moment in Lenard's descent into the ranks of the Nazis.

Lenard was not the sort one would have expected to lead the Aryan physics movement. Before the Great War, there was no trace of anti-Semitism in his interactions. He studied under Heinrich Hertz, who was half-Jewish, a fact that in no way impacted the close relationship they shared. Indeed, Lenard defended him until the end (it was his Aryan mother's influence, he would later explain in his book of great Aryan scientific heroes, that gave him his virtuous qualities).[48]

His slide into bigotry began with a personal matter. In cutting-edge science, being the first to publish and thereby having your name associated with your discovery, what is called "priority," is among the most important forms of recognition. The greatest British physicist of that age was J. J. Thomson, who is most famous for his discovery of the electron as a result of his experiments on cathode rays, which was also Lenard's area of study. Indeed, Lenard argues that Thomson knew of Lenard's work which he had sent to him, work that he claimed Thomson replicated and then published without attribution or citation. Thomson, Lenard thought, was stealing his discovery and becoming more famous and powerful for it. When he complained to Thomson, Thomson belatedly added a citation in a further publication but only to a later paper of Lenard's, assuring Thomson's claim of priority in the matter.

This incident wounded Lenard deeply and he developed a burning anger toward Thomson and also toward all of British science and the society from which it came. Lenard took British society to celebrate selfishness, encourage individualism, and thus immorality. When the war broke out and Britain was the enemy, this hatred deepened as his nationalism flowered.

This nationalism would be offended by Einstein's politics and rise to stardom, a rise that occurred in part because of the work of Eddington, a Brit. To top it all, Einstein's demeanor struck Lenard as arrogant, indicative of someone who put self-aggrandizement before nation, something that made him more British than German. So, when Lenard's conservative government was replaced with a British-style parliament ruled by a coalition of Einstein's party, Lenard's politics, his nationalism, and his bigotry became thoroughly entangled.

Further negative consequences came from Lenard's actions as a German patriot. When the Kaiser needed funds to fight the good war, Lenard dutifully bought government bonds with all of his gold, bonds that under Weimar—which he saw as a Jewish government—became worthless because of the inflation from the payment of reparations. He was bankrupted, his life savings stolen by what he saw as a corrupt Jewish administration of which Einstein was one symbol and Walther Rathenau another. When Rathenau, an object of scorn from Lenard, was assassinated, Lenard refused to lower the flag outside his institute, an action that led to his being publicly assaulted and humiliated by a gathering pro-Weimar mob.[49] At this point, Lenard's prejudice turned from personal to racial and led him to become one of the two driving forces behind the Aryan physics movement.

Lenard was valuable because he was a Nobel laureate willing to embrace the movement. His coconspirator was Johannes Stark, also a Nobel Prize winner—his in 1919 for discovering that spectral lines, the wavelengths of light given off by excited gasses, could be split by electric fields. This splitting was later explained by the quantized model of the atom, the starting point of quantum mechanics, a model and a theory that Stark flatly rejected.

Like Lenard, Stark was not predisposed to anti-Semitism or to dislike Einstein. When he was promoted to full professor at the University of Aachen in 1909, he invited Einstein to come with him as his assistant professor, a real honor. Einstein politely refused as he was in the process of securing his own position back at the ETH, his alma mater.[50]

No, Stark did not harbor ill feelings toward Einstein. In fact, Stark, who was the editor of the prestigious *Journal of Radioactivity and Electronics*, had previously invited Einstein to write a review article explaining the theory of relativity and all of the research that it spawned in the subsequent several years. Stark admired the theory, and this article helped cement Einstein's dissatisfaction with his initial theory of special relativity. It became clear to Einstein in working on this article that, because the theory could not account for acceleration or gravitation, it needed a complete overhaul, a new, more general theory would need developing. Consequently, the better part of a decade would pass before he would emerge with the general theory of relativity.

Stark, unlike Lenard, was a thoroughgoing German conservative from day one, but we can point to several personal issues that do predate Stark's move from physical objections to relativity and quantum mechanics to racial ones. The first is a priority battle with Einstein himself. Stark and Einstein independently worked out a result now referred to as the Stark-Einstein law, according to which the energy from absorbed electrons gives rise to chemical reactions. Stark contended that he was the rightful discoverer and, when Einstein dismissed Stark's contention claiming to be above such trivial matters as priority, Stark was insulted, indignantly arguing that his work was first and simpler, and thereby better.[51]

The second matter that enraged Stark was his having been passed over for a promotion three years following this spat with Einstein. The German universities were run by a government bureaucracy. Positions at a particular university were not filled by that school, but by a government education minister in Berlin. As such, every retirement created waves of politics throughout the entire university system as people jockeyed for positions for themselves and their students. When a position for a professor opened up at Göttingen, a very good job, Stark thought that he was first in line. But Peter Debye, a Dutchman who never took German citizenship and who was a student of Arnold Sommerfeld, a physicist at Munich and a close associate of Einstein, got the job instead. Incensed, Stark railed against the powerful "Jewish and pro-Semitic circle" of which he identified Sommerfeld as its "enterprising business manager,"[52] a slur dripping with anti-Semitism. Instead of Göttingen, Stark ended up at Greifswald, a refuge for nationalists, where he involved himself with right-wing politics that would pave the way for his association with the National Socialists and the Aryan physics movement.

WHILE THE NOTION OF "JEWISH SCIENCE" evolved in part from widespread anti-Semitism, especially as concerns the general opposition to modernism in all its forms, and in part from personal animosity toward Albert Einstein, the desire to reclaim science on their own terms was real for German conservatives. They contrasted what they considered the inferior Jewish approach to science with "Aryan science," a view that openly accounted for the place of race in scientific investigation and, they contended, proved itself more fruitful and produced better science.

Aryan science had a foundational metaphysical view and a preferred methodology that lived together somewhat uncomfortably. For the Aryan scientist, the universe was organic, incapable of being reduced to the deterministic, mechanical interactions of material particles. It was guided by a World Spirit whose ultimate nature was shrouded in mystery, inaccessible to human reason. But while there was a fundamental unknowable substructure to reality, its workings were clear to the careful eye of the German scientist. Painstaking observation and subtle experiments were the hallmark of good science, real science, Aryan science. Science did not start with theory, with mathematical symbols scrawled on a chalkboard, only later to be confirmed or disconfirmed by the experimenter subordinated to his master, the theoretician. No, it was the attentive observer who is the bearer of knowledge from the universe, who understands when nature speaks to him, just as Wagner's Siegfried understood the birds.

While this seems quite a strange and fanciful approach to science, there was, in fact, over a century of scientific advancement that Aryan scientific theorists cited in support of it. This version of the scientific method began from a place of unscientific racism, led to actual scientific progress, only then to descend again into pseudoscientific bigotry.

Immanuel Kant's anthropology begins with the claim that there are four distinct human races: Whites, Negroes, Hunnics, and Hindus.[53] Each of these had different "natural dispositions," that is, properties that are particular to each race. Some of these are observable, like skin color, while others dealt with intellectual capacity. Within races, there are significant differences by nationality that manifest themselves in terms of national character. We can distinguish the populations of different countries, he argued, through differences in their aesthetic and moral sensibilities. For example, we see that the French greatly appreciate the beautiful, that which brings pleasure in observation, while the British are able to partake of the sublime, that which is awe inspiring and overwhelming. The Germans, however, are able to experience both and thereby overcome the weakness and imbalance that come from overindulgence in one or the other. Africans, he argued, were in touch with neither and had feelings barely more complex than those of animals. As such, we can rank peoples in a hierarchical fashion according to depth of mind from the superior Germans on down.

Kant's student Johann Herder disagreed with his basic premise, contending that skin color and other outward features were not sufficient to distinguish people and that there was not a static linear ranking of cultures.[54] Rather, Herder argues, we see a universal advancement among them, some progressing more rapidly than others, but all guided through the same historical process toward a perfect end state according to a divine plan.

It is from Herder that Hegel was inspired to develop his own system in which this historical unfolding involves the self-realization of a World Spirit, *Geist*, through stepwise oppositions. According to this view, the whole is developing and the individual is to be seen as a mere cog in the larger metaphysical mechanism. This notion of the subjugation of the individual to the good of the whole, especially to the interest of the state, would become a cornerstone of the German picture of the workings of the world.

The sort of historical progress through synthesis we find in Hegel's worldview can also be seen in the works of Johann Wolfgang von Goethe, the celebrated German Romantic thinker. He is best known for his play *Faust* in which Mephistopheles wagers God that he can take Faust, a philosopher, away from his noble pursuits. Getting Faust to sign a contract in blood, Mephistopheles promises Faust that he will be his slave in life if Faust will be his in death. With Mephistopheles' help, Faust seduces Gretchen, his love, only to have her die in a prison from which even Mephistopheles cannot free her. In the second part of the drama, Faust moves from his pursuing his own interests to those of the state, serving the Kaiser. Still he uses Mephistopheles for his own ends, acquiring power. But when he is literally blinded by his own greed, he finally comes to see that his true will is to serve the state, the people, the *Volk*. At this point, Faust dies and is saved, freeing himself from the contract to serve Mephistopheles and admitted to heaven. Again, we have the notion of suppressing the will of the individual to the greater national cause. This concept would remain among the German Romantic objections to the Enlightenment political and economic systems of the English and French in which rational self-interest is seen as the key to social, economic, and political progress.

We see these themes emerging in Goethe's botany just as in his drama. The world was at root inscrutable, based on the mystery of the World Spirit, but through observation and the use of intuition (as opposed to reason, especially mathematical reason), one could discern the unity of Creation. The observed similarities in leaves both of plants of the same species and across species entails that they are all imperfect representations of a basic plant, an archetypal plant, an Ur-plant, which is the perfect form of the plant.[55] Everything we observe comes from such *Urphänomen*, or archetypal forms, but is taken away from them, differentiated by polarities—attraction and repulsion, expansion and contraction—that affect individuals. As with Hegel, to Goethe the true insight into reality comes from abstracting away from the individual to understand the nature of the general. It is in seeing the big picture that the unfolding of the World Spirit may be glimpsed.

Lorentz Oken expanded on Goethe's work and moved the search to zoology.[56] He looked for the archetypal vertebrate, the basic form from which all higher animals emerged. In this way, he sees man as the ultimate product of a process that starts with more basic forms and leads up to us. But as in Hegel's view, where each new phase is born of the last, so too in biology each new step in the process maintains within it traces of the elements that led up to it. And so each person is a microcosm of the entire universe, containing aspects of the entire history of creation that led up to the emergence of our species. Careful observations of our self-directed changes would offer glimpses of that deeper process and the World Spirit that gives rise to it.

And so, we see incredibly meticulous research in embryology coming out of Germany in the nineteenth century.[57] The development of embryos from the smallest cells is painstakingly documented in species after species by the likes of Friedrich Kielmeyer, Caspar Friedrich Wolff, Martin Rathke, Ernst von Baer, and Johann Friedrich Meckel. Meckel was the one who reported that the earliest stages of divergent species resemble each other. For example, he contended that bird embryos go through a phase in which they have gill slits like fish. This was summed up in Ernst Haeckel's law of embryonic recapitulation, that "ontogeny recapitulates phylogeny": we can observe the history of a species's evolution in the development of its individual members.[58]

The advances of German embryologists during the entire span of the nineteenth century provided the Aryan science advocates of the early twentieth century with a deep insight into what they considered the proper scientific method—begin with careful and thorough observation and look for underlying archetypal structures.[59] Indeed, a similar approach, when applied to physical phenomena, had given the field great advances as well. Both of the major figures leading the Aryan physics movement, Lenard and Stark, were named Nobel laureates for the discoveries that came from their slow and rigorous observations of complex phenomena. They were rewarded for doing physics the proper way, the Aryan way, starting by connecting themselves to the world, observing what it does, and then allowing it to speak to them. In this way, they could gain the insights that can be gleaned from reaching closer and closer to the mysterious World Spirit that guides the world and will forever remain beyond our grasp.

Like Hegel's history and Goethe's botany, Lenard's brand of Aryan physics contends that the basic structure of the world is based upon Ur-phenomena unfolding historically through polarities. All physical phenomena can be grouped into two classes: those that deal with material things (mechanics and gravitation) and those that deal with states of the luminiferous aether (electricity, magnetism, and optics). From these, we can, in the style of Goethe, find a more basic Ur-phenomenon in energy that connects the two.

The history of physics, to Lenard and Stark, is the story of the discovery of the workings of phenomena in these two areas by scientific heroes, great men of Aryan descent who through careful observation and intuition have uncovered them. But this fruitful and enlightening approach to science, they argued, was endangered by the rise of "Jewish science." While there was never an axiomatic description of what exactly was meant by "Jewish science," we can see the outline of it from both the contrasting notion of Aryan science and the general notions attached to the term "Jewish."

IN THE BEGINNING OF THE CHAPTER, we outlined three aspects of what anti-Semites in vogue in Einstein's time perceived as the biggest problems of Jews: they were unhealthy, disloyal, and modern. In many ways, the sense of Jews as modernist was of primary concern to their detractors. Recall that this modernism had three elements: (1) removing the world as it is perceived through rigorous observation and replacing that observation with abstract notions, symbols, and empty formalism, (2) being self-referential rather than world-oriented, and (3) introducing unnecessary complexity that distracts the thinker from the real world. We will find all of these charges leveled at "Jewish science" in the writings and speeches of the Aryan physics advocates.

In his address inaugurating the Philipp Lenard Institute for Physics in Heidelberg in 1935, Stark decried the formalistic nature of the "Jewish approach" to science:

> Certainly all scientists concur in the desire to acquire new knowledge or indeed to make great discoveries. But they differ greatly in their choice of the way to arrive at this goal; and these days usually go astray. A large group of people, primarily in physics, believe that to be able to arrive at results, or at least to come up with impressive articles or even with sensational formulations, they must produce a mathematically lavishly dressed theory or base their work on the formulas of such theories. This type of approach is consistent with the Jewish peculiarity of making their own opinions, their own desires and advantage into the measure of all things and thus scientific knowledge as well. Jewish physics, which has come about in this way in the last three decades and which has been practiced and propagated by Jews as well as by their non-Jewish students and emulators, has logically also found its high priest in a Jew, in Einstein. Jewish propaganda has tried to portray him as the greatest scientist

> of all time. However, Einstein's relativity theories are basically nothing more than an accumulation of artificial formulas based upon arbitrary definitions and transformations of the space and time coordinates . . . Despite the accumulation of mountains of this kind of theoretical literature, however, it has contributed no important new knowledge of actual fact in physics. This could not have been otherwise; since its point of departure, formalistic human opinion, was false. Jewish formalism in science must be rejected under all circumstances.[60]

The formalistic approach to physics, the Aryan physicists argue, leads to theories that are not true because they agree with prediscovered empirical facts of the natural world, but because they work with each other in their odd symbolic language games, thus following two of the "rules": being formalistic and self-referential at the same time. In a public spat with Werner Heisenberg, who spoke up to defend Einstein and his own work in quantum mechanics against the Aryan physics movement's charges, Stark shot back, accusing the new approach of using theory and not reality as the measuring stick for scientific success:

> In his article Heisenberg continues to advocate the fundamental attitude of Jewish physics even today. Indeed, he even expects that the young Germans adopt this basic attitude and take Einstein and his comrades as their models in science. He sees "systems of concepts" as the source of physical knowledge as well as its goal and speaks of a victory of the new conceptual system.[61]

The argument is that Heisenberg's commitment to these conceptual systems is a neglect of the world itself, which ought to be the sole arbiter of scientific success. Stark saw Heisenberg's emphasis on abstract concepts rather than concrete scientific facts as a Jewish fallacy.

On the one hand, Heisenberg's thought was indeed enmeshed within the systems of concepts found in relativity and the burgeoning quantum theory that he had a major hand in shaping. He was one of the inner circle of the new physics community. But Heisenberg was comfortable taking on Stark because his own nationalistic credentials should have been seen as beyond reproach. He had been a member of the White Knights, a right-wing youth movement that later folded into the Hitler Youth, and Heisenberg himself had helped in a very small way the paramilitary support for the overthrow of the communist uprising in Bavaria.[62] So trusted a nationalist was Heisenberg that he was tapped to head the Reich's nuclear weapons development program, a fact that worried Einstein and fellow physicist Leo Szillard to the extent that they sent their famous letter to President Franklin Roosevelt urging the creation of what was to become the Manhattan Project.

But though Heisenberg himself was a loyal German, his physics was a different matter. Contemporary historians of science like Paul Forman argued that the non-deterministic, non-causal elements of Heisenberg's work in quantum mechanics were the result of his mind's having been conditioned by concepts implicit in the German romantic worldview of the time. But for the Aryan physicist the association with modernism and Einstein was enough to poison the project completely.[63]

Lenard then takes up the third aspect of Jewish modern science, unnecessary complexity, in the foreword to his multivolume textbook *German Physics*, contending that Jewish science is that which in its self-referential use of symbolic formalism obfuscates science in a fashion designed to bamboozle.

> To characterize [Jewish science] briefly, let me best refer you simply to the activities of its undoubtedly most prominent representative, to the unquestionably pure-blooded Jew A. Einstein. His "relativity theories" attempted to transform and dominate the whole of physics; but they have now already completely played themselves out against reality. Apparently, they were never even intended to be true. The Jew conspicuously lacks any understanding of truth beyond a mere superficial agreement with reality, which is independent of human thought. This is in contrast to the Aryan scientist's drive, which is as obstinate as it is serious in its quest for truth. The Jew has no noticeable capacity to grasp reality in any other form than as it appears in human activity.[64]

Jewish science is thus, in Weyland's words, "scientific Dadaism," an absurd, symbolic game designed to baffle and confuse rather than to carefully and reverently uncover the nature of reality in a fashion that makes it clear to all.[65] So, relativity is Jewish science because it is fully modern using all three criteria established earlier.

THE SECOND CONNOTATION ATTACHED TO THE TERM "Jewish" is disloyal, something implicit in the notion of internationalist, and the advocates of Aryan science saw this explicitly in Jewish science:

> In reality, as with everything that man creates, science is determined by race or by blood. It can seem to be international when universally valid scientific results are wrongly traced to a common origin or when it is not acknowledged that science supplied by peoples of different countries could only have been produced because and to the extent that other peoples are or were likewise of a predominantly Nordic racial mix. Nations of different racial mixes practice science differently.
>
> Yet, no people has ever embarked on scientific research without basing themselves on the fertile ground of already existing Aryan achievements . . . Jewish physics has developed and become prevalent, which has only rarely been recognized until now, because literature is usually categorized according to the language in which it is written. Jews are everywhere; and whoever still contends that science is international today clearly means unconsciously Jewish science, which is of course, similar to the Jews everywhere and is everywhere the same.[66]

The idea that science is objective and therefore universal was seen as a Jewish lie designed to help their pernicious approach infiltrate physics despite its superior Nordic roots.

To show this, Lenard produced a popular book, *Great Men of Science,* which shows that science was led forward by great individuals of Aryan descent.[67] At first glance, it is just another innocuous popular introduction to the history of science, highlighting the great advances and the thinkers that developed them. But implicit in it is the Romantic great man theory in opposition to the Enlightenment rationalist picture. Only figures conforming to the Aryan science racial image are selected, although he includes his beloved mentor Hertz, even disclosing that "Heinrich Hertz was the son of a lawyer and senator in Hamburg, and partly of Jewish blood."[68] But he does so only after he makes clear that Hertz was doing science in the Aryan way, using intuition to fuel his discoveries and then using theory as Stark put it, as "an arithmetical and representational aid."[69]

> Hertz was the first who not only understood the equations, and knew how to deal with them mathematically when necessary, but also saw the structure of ideas upon which they were based by their originator, and understood how to move about in it. The equations are, so to speak, merely ground plans of this structure, and are far from being actual inhabitable apartments; the latter can only be produced by the architect, who knows how to grasp ideas which have been put into the ground plans.[70]

Indeed, in later editions of the book, Lenard separates Hertz's earlier discoveries from his later more theoretical work, arguing that the former is far superior and the result of his Aryan mother, while the later is the malicious influence of his father's lineage.[71]

Oddly, as much as Lenard railed against Einstein's work, his book ends with the great discovery of the relation between energy and mass, $E = mc^2$, the equation well known by the great, the immortal, the world famous . . . Friedrich Hasenöhrl. Lenard found reference to the relation in Hasenöhrl's work and seized on it as a way to squeeze Einstein completely out of the history of science. He labels it an "important and highly remarkable result"[72] that is not yet empirically nailed down—"it needs continual further proof in its application, until it is rendered sufficiently secure, and the limits of its validity are known."[73] The only reference to Einstein is a subtle back-handed one: "The applications of this idea have already progressed very far to-day, although almost entirely in the names of other people."[74]

Hasenöhrl was useful in two ways for Lenard's rhetorical purpose. First, not only was he not Albert Einstein, but he was not even Jewish. To the contrary, he was an Aryan scientist who died fighting in the Great War, a brave, patriotic Austrian martyr for the cause.

> Then the war broke out, and Hasenöhrl immediately volunteered. He was everywhere at the front, first at the defense of Przemysl, then in the Tyrolese mountains, which he knew and greatly loved. After a bullet wound, which was more or less healed, he again went to the front, and fell in the second year of the war at Vielgereuth, only forty-one years old. He was of a simple, kindly and modest nature. Only when it was a matter of self-sacrifice was he to be found in the front ranks; when it was a matter of gaining rewards, he was in the background, even when he was the person most concerned, as we see in the most astonishing manner in his writing.[75]

So, unlike Einstein, this model Aryan was bravely standing up for the defense of his nation, always modestly putting country before self, not seeking self-aggrandizement.

Attributing the theory to Hasenöhrl was also useful for Lenard because it allowed for the use of results in ways that Einstein clearly objected to. Einstein held the theory of relativity to have rid physics of the luminiferous aether, something that Lenard took to be a necessary part of the universe. But Lenard reinterprets $E = mc^2$ in such a way that, instead of undermining his Germanic view of the universe, the equation is taken to bolster it. By showing that mass is equivalent to energy, we can unite the two sides of physics into a single Ur-phenomenon. We can take the Ur-aether, the true subterranean space, to join the seemingly dichotomous worlds of waves, on one hand, and particles, on the other, into a single realm of physical reality. Just Hegel and Goethe's pure German thought proceeds to unify history and life according to polarities, so Lenard can follow suit in the realm of physics. So, by claiming relativity theory for the Aryans through Hasenöhrl's name, he can assert relativity not to be international physics, not to be Jewish physics, but in fact to stand in favor of Aryan physics . . . should it turn out to be true, to which he still isn't committing himself.

THE THIRD CONNOTATION OF JEWISH was unhealthy, and we see the Aryan notion of Jewish physics as partaking of, indeed infecting, German culture with Jewish unhealthiness. Hitler put it this way:

> Science, once the greatest pride of our people, today is taught by Hebrews, whose science is nothing but a means by which our national soul is being systematically poisoned and thereby bringing about the collapse of the core of our culture.[76]

Inherent in his comment is the notion that science had changed. Just as Nietzsche contended that, with respect to ethics, a change in political power led to an inversion of values from healthy, life-affirming world-directed aristocrats to life-denying, linguistic, conceptual, unhealthy Jewish values, the Aryan science advocates contended that a similar harmful revolution was under way in physics.

> Slowly, very slowly, the truth prevailed; slowly we began to become aware of the rules of physics; slowly people's conceptions of the world became that of physics, and the nation became receptive of new ideas. In the wake of the revolution in physics came theoreticians like Einstein who then strove to turn physics into a purely mathematical system of concepts. They propagated their ideas in the manner characteristic of Jews and forced them upon physicists. They tried to ridicule men who criticized this new type of "science" with the argument that their intellect just could not aspire to the lofty spheres of the Einsteinian intellect—an intellect which, says Lenard, does not consciously seek after truth.[77]

By that account, there was a slave revolt in physics, just as in ethics, and the Jews, finding a point of weakness, slipped in unnoticed and changed science. And just as Nietzsche's first inversion caused the physiological decline of the fit and virile bodies of the blond beast, so, too, this second Jewish inversion, the inversion of scientific methodology, caused the impotence of modern science.

> The Jewish spirit expresses itself in science and art in exactly the same fashion it did in morals and religion. The Jews pride themselves on having contributed, in all ages, a great number of prominent men in science . . . With all their knowledge, they lack intuition of genius and creative strength. Since Kant, we have differentiated precisely between understanding and reason. With the former we associate the ability to gather together sensuously-mediated data to form a picture and then tie it together under forms of causality. With the latter term we mean the ability to bind together all judgments of understanding into a unity. Understanding creates knowledge. If, on the other hand, reason comprehends the given, it is nevertheless spontaneously active and in the capacity of a bold, direction-giving idea, it pushes the feeler towards new discoveries. The idea of the atom, the law of conservation of energy, the ether theory—these are not things which any fool could formalize; they cannot be proven logically without further ado. They are forward-reaching attempts of creative reason, "precise sensuous phantasy," as Goethe called it, which goes hand in hand with irreproachable empirical investigation.
>
> It is now difficult to delineate with great clarity the sphere of the Jewish spirit. From time immemorial it has mastered that region of science which has been occupied with understanding. That lack of phantasy and inner searching, damned to fruitlessness by the religion and philosophy of the Jews, is reflected in science as well. Not one single creative scientific idea has sprung from the brain of a Jew, he has never been a pathfinder.[78]

"Not one creative idea has ever sprung from the brain of a Jew." Science done in the Jewish style is barren, rendered fruitless. Take that Albert Einstein.

SO, IF THIS IS WHAT THE NAZIS meant by Jewish science, a notion of science that was in large part crafted to undermine and disparage the theory of relativity in particular, does it really hit its target? By the Nazi's own concept, is the theory of relativity, in fact, Jewish science? Let's consider it point by point.

The advocates of Aryan physics contend that the theory of relativity is a further example of modernism: self-referential, putting formalism in the place of physical reality, and unnecessarily complex. The claim of self-referentiality in the theory of relativity seems to be appropriate in that like other examples of modernism: Einstein was not only doing physics but also redoing how to do physics. Just as Mahler or Tzara were creating art and simultaneously engaging in challenging what it means to make their particular form of art, so too Einstein radically revised central notions of physics like space, time, and mass and rejects the notion of the luminiferous aether. Basic assumptions underlying our understanding of the project itself were altered and repudiated for a new approach to an old practice.

But unlike other modernist movements of the time in which practitioners actively excavated foundational presuppositions to challenge, Einstein's reframing of basic physical concepts were not sought out for the sake of overturning them. Hippolyte Fizeau and Albert Michelson and Edward Morley had given us empirical results that needed to reconciled. How do we make sense of all of our observations? Can we do it in terms of the concepts we had been using? We tried. We couldn't. Einstein figured out how to do it, it just took some conceptual twists.

The irony, though, is that this rethinking of the basic notions of physics derived from the results of rigorous observation that the Aryan science advocates argue ought to be the starting point of science. You take careful measurement and then with the sort of creative insight of which Goethe wrote, you let nature tell you how it works. That is exactly what Einstein was doing. In that aspect of the Nazi's own notion, Einstein was doing Aryan science.

What of the second criterion? Is the theory just a collection of symbols, no longer relating itself to the world? A number of critics at the time argued precisely that. Oskar Kraus, for example, argued that the heart of the theory of relativity is nothing but "mathematical fictions."

> One can interpret the Lorentz contraction only as a mathematical consequence resulting from certain fictitious measuring operations; more precisely: one obtains the value of the Lorentz contraction when one figures out what the results of certain measuring operations would be. Actually, such a measurement has never been performed. Thus, the calculation of the measuring results under certain fictitious conditions is confused with actual measurement.[79]

The theory of relativity is not a new and better way of looking at the world and understanding the universe; rather, it is a cute mathematical sleight of hand.

It is certainly true that at normal speeds, the relativistic effects are not noticeable, but does this mean that the entire theory is just a game of manipulating mathematical formulae? Was Einstein removed from the real world, working only in a realm of abstractions?

No. Actually, those, like David Hilbert, who did work in the world of formalistic abstractions, the purely conceptual mathematicians, actually belittled Einstein precisely for his lack of skill in their symbolic games.

> Every boy in the streets of our mathematical Göttingen understands *more* about four-dimensional geometry *than Einstein*. Yet, despite that, *Einstein* did the work and not the mathematicians.[80]

Here, Hilbert draws a sharp distinction between Einstein and the mathematicians who do, in fact, do what Einstein was accused of. Indeed, when Einstein first examined Minkowski's geometric interpretation of the special theory of relativity, exactly the sort of thing the Nazis are pointing out as typical of the theory, he did not immediately grasp its full impact.

While Einstein was capable enough with the mathematical tools *for a physicist*, he was not someone who worked from the equations first. Rather, Einstein worked from his intuitions. Notice the style of his arguments. Einstein starts from a thought experiment, from a situation we could set up in the real world and tries to get himself in touch with what the universe would do in that case. In the debate at

Bad Nauheim, Lenard explicitly confronted Einstein on his use of thought experiments and how Einstein could rule out those that did not support his theory. Einstein responded that only those that could in principle be conducted were allowable.[81] In other words, we need to step away from the mathematics and think in terms of pictures that could be the case in the real world based on actual observations. He then made appeals to intuition exactly in the fashion that the Aryan scientists' beloved Goethe says to. Again, Einstein seems to be more of an Aryan thinker than a Jewish one by the Nazi's own lights.

This is not to say that there were not physicists at the time who certainly satisfied this condition, who elevated the formalism and really did do physics in an abstract world of symbols. Einstein's dear friend, Max Born, who also had a Jewish heritage, would be one example.[82] Born's work on the foundations of quantum mechanics was late in being recognized for its importance, in part because of its technical, formalistic nature, in part because of political maneuvering by Werner Heisenberg and Niels Bohr who were also seeking recognition in the fast-shifting world of quantum mechanics in the 1930s and 1940s.

Born, initially trained as a mathematician, was the first to realize that the proper mathematical setting for the rough results of early quantum theory was in terms of what mathematicians call "matrices." The Euclid of quantum mechanics, Born worked diligently to translate the isolated elements and results of the theory that were floating around the community into a coherent, organized mathematical and logical structure using matrix-based formalism. He had found a means of elegantly structuring the unruly mess and allowing further research to be done rigorously.

At the time, he had a brilliant, talented assistant who was also well trained in the necessary mathematical tools. Pascual Jordan worked closely with Born and, once quantum mechanics had become matrix mechanics, Jordan was able to make major contributions concerning details of playing with the formalism used in the theory. Jordan's work in the early days of quantum mechanics was the quintessential sort of "Jewish" science—purely formalistic, completely removed from observation, entirely concerned with the ways in which one should manipulate the symbols.

The irony is that Jordan was a member of the Nazi Party; and even became a storm trooper. So we have one Jew in Born whose work satisfies the criterion for Jewish science and one in Einstein whose work does not, and we have a Nazi whose work satisfies the criterion for Jewish science in Jordan and another in Lenard whose work does not. The concepts of Aryan and Jewish physics seem to have a problem here.

What about the third part of the modernism condition, that the theory of relativity is unnecessarily complex, that it tries to obscure the true nature of the observable world? There is no doubt that the four-dimensional semi-Riemannian space-time manifold with all the properties given to it by general relativity is bizarre and the results of the special theory, much less the general, seem odd and counterintuitive at the least. But, they were posited precisely because the common sense views simply couldn't account for everything we observe when we do the sort of careful observation and measurement that Lenard and Stark yearn for. They are complicated theories, yes, difficult to work in, true, but they are the simplest that gets done what physical theory needs to do, account for the observations we make of the real world. It is the universe that was getting weirder and weirder as we looked at things that went faster and faster, not just our theories about it. It would be wonderful if we lived in a simpler world like the one described by classical physics. We just don't.

What of the disloyalty aspect? Is the theory of relativity international science? On the one hand, the special theory of relativity is the result of trying to account for the experiment of two Americans (Michelson and Morley) in the work of a Dutchman (Lorentz), which when combined with the methodological insights of a German (Mach) and a Frenchman (Poincaré) led to conversations in Switzerland with his friends in the Olympia Academy from Romania (Maurice Solovine) and Italy (Michele Besso). On the other hand, Einstein does not put his theory's multicultural heritage on display. In fact, it is remarkable that in his paper he cites none of his influences at all with the exceptions of thanking Besso for "several valuable suggestions."[83]

Finally, what of the healthiness of physics following relativity theory? Stark and Lenard are correct that the theory of relativity was part of a wave of advances that radically changed physics as a process. The sort of story that Nietzsche tells in terms of an inversion with respect to morality does also take place in the physical community. Before the turn of the twentieth century, the physics of Newton and Maxwell were capable of accounting for virtually all observed phenomena, and there seemed to be good reason to think that those not yet accounted for were merely a matter of time and cleverness.

But this was wrong. A new physics was needed to account for things that are very fast, very large, very small, or radioactive. Everything was done to try to save the old physics, but all attempts failed. In its place came a pair of new theories that did the job. They were capable of accounting for the odd phenomena that occur in contexts unlike those we normally inhabit and which reduce mathematically to the old ways for the situations of the sort we usually encounter. Calculations with them reveal amazingly strange situations and phenomena that we would never have known to look for before. Absent the theory, no one would have had any reason to think to check for such possible effects, but they fall naturally out of the theories.

As such, with the positing of relativity theory and quantum mechanics, there was a role reversal in the community. Theory was no longer the handmaiden of experimentation. Physics changed. It became more mathematical, requiring mathematical tools physicists had never needed before and new concepts instead of the comfortable old ones. These new, strange theories made predictions that would not have been considered with scientists working in the realm of experiments leading the parade.

This change, of course, was deeply upsetting to experimentalists like Lenard and Stark. And well it should have been. The scientific fraternity was no longer governed by the old rules it had always had, the rules under which Lenard and Stark achieved places of renown as model practitioners. Doing physics was very suddenly a different endeavor than it had been only a generation earlier, but, of course, this change did not deprive the field of its fruitfulness. To the contrary, everything from the most abstract of the cosmological models to the most practical material sciences that emerged as a result of these advances have invigorated the field. The theories that overturned the old worldview opened up frontiers that the most creative romantic mind could never have conceived of. Physics after Einstein, of course, is still alive and kicking.

So, the Nazis crafted a notion of Jewish science with the express intention of capturing relativity theory. Did even that sense of the term fit? Yes and no, but even where yes, certainly not in the way they hoped. The Nazis' claim that relativity is Jewish science by and large fails even when we grant the Nazis' own definition.

CHAPTER FIVE

Did Relativity Influence the Jewish Intelligentsia?

The claim that Einstein's theory of relativity is "Jewish science" was a Nazi construction, but is it possible that the Jews could own it? While we ordinarily scoff at the phrase "Jewish science," we have no problem categorizing certain writers throughout history, including contemporaries of Einstein, as Jewish thinkers. There was a vibrant collection of Jewish intellectuals who were wrestling with deep and interesting problems in a fashion that certainly spoke to those beyond the Jewish community but that drew on Jewish thought both as an inspiration and as a source of insight.

Einstein was no stranger to these people. In fact, Martin Buber, perhaps the most famous of them, was a close friend of Einstein's, so it seems reasonable to ask whether the theory of relativity could be thought of as "Jewish science" in the sense that it was part of the conversation occurring among Jewish intellectuals at the time.

The phrase "Jewish intellectuals," of course, falls into exactly the problem we examined in chapter 1: who is a Jew? Of the writers at the time, which ones do we count in and which ones do we not? Let's consider two groups. The first were thinkers who were explicitly wrestling with questions about what it is to be Jewish, those who were part of the movement to reclaim and redefine Judaism. We can contrast them with the second group, prominent scholars of Jewish heritage who would in no sense have called themselves Jewish thinkers, indeed, would not have considered themselves Jewish at all. These are secular thinkers who were completely assimilated but working on deep foundational questions about what it is to have knowledge and how we live as humans.

For the explicitly Jewish thinkers, we'll see that Einstein was not at all relevant to their examination of Judaism, although they did have an influence on Einstein. For the secular thinkers, Einstein precipitated a rift, the so-called analytic and Continental divide that partitions the philosophical world to this day. It is a deep disagreement concerning the foundations and methodology of philosophy. The analytics see science as the basis for a worldview and use Einstein's theory and means of reasoning as a template for all thought. Contemporary Continental thinkers, in contrast, saw a totalizing view of science as dehumanizing our picture of the world, something that could have tragic consequences and needed opposition.

WITH ALL OF THE TRADITIONAL ANTI-SEMITIC social structures and institutions in the decades before Einstein, Jews were second-class citizens. Thus liberal politics based on Enlightenment values, according to which all humans are to be considered equal under the law, attracted the loyalty and participation of a large segment of the Jewish community.[1] Liberalism allowed Jews economic mobility, some measure of political power, and the chance to pursue fields of study like science, law, and medicine, success in which would demonstrate the equality of Jews with their fellow countrymen.

The liberal worldview is deeply atomistic in that it treats society as a collection of individuals, each of whom looks out for his or her own best self-interest. Then this in aggregate creates the maximally successful society. We bargain for ourselves in the marketplace, think for ourselves at the ballot box, we are beings each and only for ourselves. This individualization applies to everything, including religion, which becomes a matter of personal faith. German Jews, in freeing themselves from the bigotry of the old ways were in many ways legitimately emancipated, and in this profitable transformation labeled themselves as "Germans of Mosaic belief"—that is, Mosaic in the sense of "of Moses," not in the sense of a patchwork artistic creation (even if that would be an appropriate metaphor for the community).

This notion of religion as a matter of personal belief is a loaded concept. It presupposes that religion is about faith, what one takes to be ultimate truth, a state of mind, not a set of actions or rituals, a tribal notion of belonging, or a way of life. Traditionally, Judaism is about doing, not believing. Jews are commanded to act in certain ways, and it is on the basis of carrying out these actions that one is Jewish. The emphasis on the mind, belief, and faith comes from the Greeks and embeds itself in Christianity because of the influence of Greek thought on early Christian doctrine. This emphasis only enters Judaism in the Middle Ages, once thinkers like Maimonides tried to reconcile Western philosophical ideas with traditional Judaism, which had made following the commandments and properly executing the rituals paramount. The liberal individualistic approach in which religions are seen as "faith traditions" and membership connected with the internal beliefs of individuals, inadvertently redefined religion in general in line with an implicitly Christian conception. Liberalism, thereby inadvertently redefined Judaism in a way that was not inherently Jewish.

Even though the history of liberalism is rooted in Protestantism, its advance left it thoroughly shot through with a scientific worldview. Since liberalism as a whole was so successful in advancing the lot of individual Jews in terms of acquiring wealth, knowledge, status, and rights, Jews applied it to Judaism itself. Anthropological, historical, and sociological scholarship replaced rabbinic training for many as ways to be Jewish thinkers. Jews were keen to show that being Jewish was not inconsistent with being a modern, urbane citizen. Jews were just like everyone else, and it was only right to treat them that way.

But while Jews were individually flourishing from this new approach, the Jewish community languished. Left on their own, looking out for themselves in this atomistic way of life, many Jews—like

Einstein's parents—sent their children to schools run by other faith traditions. These children soaked up what German culture had to offer and realized that they had much to gain by assimilating as completely as possible. But while being freed from the ghetto came with benefits, it also exacted the cost of alienating Jews from each other and their own traditions, undermining the coherence of what had been a strong community.

The rise of Kaiser Wilhelm and the cultural shift giving place to the nationalists during and after the Great War made it clear to Jews throughout Germany that the traditional anti-Semitism they faced had not dissolved. Jews realized that they were not, and likely would never be seen as just like everyone else. And thus they found themselves caught in the same antimodernist quandary that the non-Jewish Germans faced: if we are not just a bunch of rational beings with equal rights, then who are we? As the Christian Germans shook off liberalism and tried in their own way to define what it is to be an authentic German—definitions that clearly and explicitly ruled out German Jews—so, too, the German Jews needed to rediscover or redefine what it meant to be a Jew, the old ways having been discarded in the false hope of complete assimilation. Germans were forming Wagner Societies to try to rekindle pride in what they came to construct as traditional Aryan ways that they believed had been wrongly erased from the culture in its short-sighted approach to modernizing. Similarly, the old ways among Jews had been rejected as embarrassingly primitive superstitions that no modern citizen would cling to, replaced with what some held to be a sanitized scientific objective worldview. But this left Jews without Judaism. For the same reasons and in the same way that non-Jewish Germans reformulated a notion of German authenticity without the purely rationalistic picture of liberal modernism in a way that would ultimately give rise to the horrors of World War II, so, too, the Jews of Germany sought to reformulate what it was to be an authentic Jew in a way that rejected modern, liberal rationalism.

One of the founding fathers of sociology, Ferdinand Tönnies, published his famous and influential book, *Community and Society*, at this time.[2] Its central theme is distinguishing the notion of community, a group that is unified by some set of common attributes, from the notion of a society, a group defined by its internal diversity. Premodern culture, Tönnies argues, is dominated by the notion of community. We once lived among people who were like us and who learned the ways according to which we lived our lives. But modern life is based on the concept of society. Tönnies felt that modern individuals lived in more cosmopolitan, industrialized settings in which people strove to express their individuality. Tönnies was not a conservative yearning for the good, old days of community but rather a social scientist trying to make sense of the radical transformation of culture that he—and everyone else—was seeing. Indeed, he argued that we could never have a fully communal life, because every community gives rise to societies (no matter how homogenous we think we are, we will inevitably find some characteristic to cause us to break into factions), and every society will form a community (inevitably finding external common enemies, real or imagined, that bind us together in a joint struggle).

Tönnies's concept of community lost captured the underlying goals of both the non-Jewish and Jewish Germans alike, who thought it a task of the utmost urgency to reconstitute community. This meant figuring out what one's community was based on and enumerating the commonalities that bound the members together. Whatever they were, they would not be the sort of universal attributes the prorational, proscientific modernists put forward, after all, those were the ideas that undermined the sense of community in the first place.

And so with a growing sense of alarm that modern Jews knew little, if anything at all, about Judaism, a sense of urgency arose around the task of reestablishing a sense of authenticity within the Jewish community, of reconnecting with something that had been lost and in doing so uniting the Jews of the West in a fashion that recharged Judaism. The movement of assimilation had found its counterbalance in a movement of dissimilation. The problem with this attempt to reconnect with Judaism was that no one had enunciated a clear answer to the central question: what exactly *is* Judaism? Proposed answers can be broken down into three main approaches: Judaism as a religion, Judaism as a tradition, and Judaism as a nation.[3]

THE NINETEENTH CENTURY SAW A SPLINTERING of the Jewish religious community. With the modernist move toward assimilation and the social/political/economic good that could come to Jews when they inserted themselves more fully and naturally within mainstream culture, a movement arose to reinterpret Judaism in line with these Enlightenment-based beliefs. The goal was not to cause a schism but rather to reorient the religious direction of the entire spiritual community, to update it with a more modern sensibility. Reformers argued that there were elements that had developed in both community custom and rabbinic interpretation that no longer connected with the way of being in modern life. Times had changed, humanity advanced, knowledge grown, science progressed, and, if it were to remain relevant, Judaism too would have to adapt to this new world.

The Torah is a timeless and sacred text, they allowed, but the Talmud is a historical document—understandings and interpretations tied to cultural contexts long left behind—and not to be subject to idolatry.

> The Talmud was not an infallible authority, that the rites and ceremonies of the religion had to be subjected to research, and if found to be subversive to the spirit of true religion rather than helpful, to be changed or discarded; ceremonies that furthered the religious life in one generation might be a drawback to another.[4]

To the traditionalists, this is tantamount to heresy. In the words of Rabbi Tiktin,

> whoever disregards any command or prohibition of the Talmud must be considered an unbeliever and as standing without the pale of Judaism, and is therefore an untrustworthy witness.[5]

The push by reformers, the traditionalists claimed, was tearing the heart from Judaism. So a countermovement formed, giving rise to

Orthodox Judaism as an attempt to reverse the modernizing momentum of Reform Judaism. Judaism is not historical, they contended; it is not a matter of fad or fashion. You don't have to see what the rabbis are wearing this fall—basic black always goes with Judaism. There is a single strand reaching back to Sinai and deviation is the same as outright rejection. Judaism is, on this view, a faith, a religious set of truths to be believed and upheld.

The battle within the religious community was vitriolic for generations, often getting reduced to personal attacks and the cheapest, ugliest brands of dirty politics.[6] In larger cities where the community was sufficiently large and where Liberal Judaism was its most influential, separate synagogues and yeshivas would ultimately be established. In smaller, more rural areas, where such division was not logistically possible and where a more Orthodox approach reigned, informal means of coexistence would have to be found.[7]

The quest to mediate between the need to connect with the modernized Jew living a comfortable bourgeois life in the urban, industrialized world demanding equal rights as a citizen and the desire to retain that which was part of the living tradition, that which seemed to be the essence of what had been handed down generation to generation in a sacred trust, gave rise to a third approach. The Historical School, which would later become known as Conservative Judaism, agreed with the Reformers that Judaism must be seen as historical, something that develops over time. But where the reformers saw this as license to change Judaism in ways that fit the modern lifestyle, the Historical School saw the research into the history and anthropology of Judaism as a means of mining the true essence of the religion, of uncovering that which is authentic and, like the Orthodox, held that it must be maintained. It combined the sort of modern scholarship of the Reform movement with the adherence to tradition of the Orthodox. This required a separation of acting from belief.

> The ambivalence of Conservative Judaism, speaking in part for intellectuals deeply loyal to the received way of life, but profoundly dubious of the inherited world view, came to full expression in the odd slogan, "Eat kosher, but think *traif*," meaning that people should keep the rules of

the holy way of life, but ignore the convictions that made sense of them. *Orthopraxy* is the word that denotes correct action and unfettered belief, as against orthodoxy, or right doctrine. Conservative Judaism . . . could thus be classified as an orthopraxy Judaism defined through works or practices, not through doctrine.[8]

The Historical approach was designed as a middle path that accepted the notion of historical development but did not break ties with that development. We may not believe what Jews in ancient times believed, but that doesn't mean we shouldn't do what Jews in ancient times did.

In all three of these modern Judaisms, we find the same wrestling with the question of authenticity, of what it means to be Jewish. While much of this debate predates Einstein, it certainly continued through his lifetime. Does his thought come into play in any of this? No. For the religious community, Einstein represents the sort of secularized modernism that was the precisely the problem. Abraham Heschel, a rabbi a generation younger than Einstein who rose to significant prominence, contended that Einstein's worldview was a threat to Jews and Judaism itself.[9]

Convinced "of the existing order of *all* natural events," Einstein dismissed the good Lord from His dwelling place in the universe. That is rather surprising. We didn't realize until now that "*all* events"—including history, the life of the soul, and the realm of the intellect (the arts, law, ethics, etc.)—took place within only and nothing but a natural and rational order. Historians, psychologists, sociologists and artists, so far, have observed only very little of the "natural order of *all* events," only those of nature and ever present rationality . . .

As for the shimmering palace of the Good, the True, and the Beautiful, which Einstein attempts to erect out of the natural sciences, it is preceded by an architectural design from Haeckel's pen. Now really, this is a castle built in the air, for one cannot conjure up these three pillars, and particularly the Good, on the foundation of nature. It is the privilege of natural science to be value free. In itself, it cannot be hindered from inventing poisonous gas or dive bombers; and rationalism is powerless

> once "the magnificent blond beast" institutes slave morality lusting after "prey and victory," taking up weapons in order to subjugate inferior races. After all, nature—toward which Einstein demands a humble attitude—is governed by the brutal law of the strongest. Were nature to become the ultimate source of knowledge, we would have to accept bestiality and fatalism. But then there would be no freedom, no truth, no science. Just try to think through this awful chain to its final consequences. That is the reality![10]

To Heschel, Einstein's rationalistic approach was responsible for the intellectual foundations that led to the destruction of European Jewry. The religious approach must step in to save Judaism from the forces Einstein represents.

BUT, AS THE TASK OF RECAPTURING Judaism was seen as more and more crucial within the community, the shul was playing less and less of a role in it. The rediscovery of authenticity for Jews was therefore also being proposed in forms that were separate from the overtly religious. Judaism should be understood in terms of community membership, the claim went, a community that had roots, ways, and traditions that had been surrendered and needed to be recovered. It was not about reconstructing the Temple but instead about reconstructing the community.

Leading this movement was Franz Rosenzweig.[11] A model of liberal assimilation, he had received his doctoral degree for a dissertation on Hegel and, like other members of his extended family, planned to convert. But before he split from Judaism, he thought it only proper to investigate the tradition he was to deny, so that the rejection would be authentic and complete. Contrary to plan, his studies in Judaism led him to embrace his heritage and to champion its scholarship and dissemination. While in the trenches of the Great War, he wrote his masterpiece, *The Star of Redemption*, but it was after the war that he was to have his greatest impact.

In August of 1920, Franz Rosenzweig opened the Frankfurt Free Jewish Lehrhaus. *Lehrhaus* is the German translation of *beth midrash*, or house of learning. By adopting the German instead of Hebrew,

Rosenzweig was sending a message. This was to be a new kind of place. Rejecting the traditional pedagogy of a learned rabbi lecturing down to a less knowledgeable audience, Rosenzweig had a better idea. Learning is not a sage pouring the contents of his head into the heads of his students; it is not absorbing the contents of a book. Learning happens when people appropriate knowledge, own it, making it a part of themselves.

> Hebrew, knowing no word for "reading" that does not mean "learning" as well, has given this, the secret of all literature, away. For it is a secret, though quite an open one, to these times of ours—obsessed and suffocated as they are by education—that books exist only to transmit that which is between the achieved and developing, that which exists today, at this moment—life itself—needs no books. If I myself exist, why ask for something to "educate" me?[12]

For Rosenzweig, knowledge of Judaism had been the result of an autodidactic process; he did it for himself, to himself, in a way that no one else could have done. He would create a space for everyone in the community to engage in the same sort of self-discovery.

His *Lehrhaus* did have some lectures, but mostly small study groups. The lecturers and group leaders were, by in large, not rabbis, but scholars from the sciences and humanities who themselves had rediscovered their Judaism and whose personal enthusiasm and dedication would serve as model for the rest and as a source of energy for the group.[13] The only thing more powerful than the zeal of the convert is that of the revert. If the leaders did not have an encyclopedic background in Judaism, so be it. The entire endeavor would be an exploration for all and the thrill of discovery would lead everyone onward.

The *Lehrhaus* was a place for modern Jews to reconnect with themselves as Jews without having to necessarily surrender being modern. The idea was to bring in Jews of every socioeconomic status and level of educational attainment and give them a place to actively engage with their background and the other members of their community—young and old, rich and poor, male and female. But,

Rosenzweig always said, never go a day without a line of Hebrew, as he and his wife taught Hebrew language classes.

The term "free" in the name "The Frankfurt Free Jewish *Lehrhaus*" did not mean that there was no fee associated with attending. Rather, it indicated that it was not aligned with one of the movements fracturing the religious community. All Jews were welcome. Indeed, the idea was to have a place to heal the wounds and reintegrate secular and observant Jews of all stripes. The *Lehrhaus* did have to support itself financially, and Rosenzweig thought it a good idea to bring in a marquee name, a superstar of the Jewish world, and so he brought in Martin Buber.

Buber had long been well known in the Jewish communities of Germany as a leading light of the new "Jewish Renaissance," a term he himself had coined. If the rationalism of Enlightenment modernism was to blame for the contemporary alienation from Judaism, then the reversal of the effect would require the opposite of the cause. Classical irrationalism would be the cure. In the search for what it meant to be an authentic Jew, Buber was among those who turned their gaze eastward. The culture of Western Europe had robbed Jews of their sense of self, so it would be a rejection of reason and a return to the ways of the Jews of Eastern and Central Europe. Buber, like Tönnies, thought that there was a different way of life in unified communities. So he examined the Hasidim, a group of Jews who had a much more mystical sense of the religion and whose lives were portrayed as simpler, more devout, and more joyful. Buber's book, *Tales of the Hasidim*, became a touchstone for the new movement. The mysticism, the miraculous, the *bubbe meises* (old wives' tales), and stories of golems and dibbuks were part and parcel of the Jewish sensibility that had been stripped by modern life and its objective, empirical way of being. To recapture Judaism for its lost Western European tribe, the old elements had to be reintroduced.

Just as Germans adopted the brave Siegfried as their icon of German identity, the Jews of Germany took a romanticized picture of the Eastern European village Jew as emblematic of pure, authentic Judaism. Much like contemporary American politicians and commentators take "Joe Sixpack," a white, Christian, working-class

Midwesterner as the "true American," so, too, the cultured Jews of Germany looked to their non-urban, less-well-off cousins as the embodiment of the "true Jew." And so romantic art began to be produced featuring village dwellers and old rabbis with long white beards to capture this sense.[14] Kabala and Zohar, the foundational works of Jewish mysticism, long ignored as remnants of an embarrassingly immature past, came to be reexamined with interest, both anthropological and spiritual. It was all about re-creating the old ways (or at least romanticized versions of what they were thought to have been) that had bound together Jews as a community in the past, and, they hoped, would do so again.

In this look eastward and backward in time for authenticity, Einstein's new theories of space and time looking forward and upward held no interest. Einstein was entirely irrelevant. The theory of relativity did not exist in their intellectual universe, even if Albert Einstein, the man, did. Recall that Buber was no stranger to Einstein and that the two were friends. Buber's maid, Hedwig Straub, recalled that "Albert Einstein was a regular visitor at the Buber house."[15] By many personal accounts of time spent with Einstein, one could not enjoy his company for any significant amount of time without having the theory of relativity explained to you. So, it would seem quite unusual for Einstein to have been a regular visitor at the Buber home and Buber not to have received a personal introduction to Einstein's great work. Yet, any trace of influence from Einstein is not to be found in Buber's thought or that or any others who were seeking to recapture the communal notion of Judaism.

The case is different when we look at the Einstein-Buber relationship in the other direction. Buber clearly influenced Einstein—not in his physical reasoning but in his understanding of Judaism. Recall Einstein's two pronged definition of Judaism from his speech "Is There a Jewish Point of View?" according to which being a Jew meant (1) having a sense of awe and wonder and (2) treating the relationship with other as a source of moral obligation. We see Einstein clearly bringing his view of Judaism in line with Buber's in pivotal ways.

Buber's view of the Jewish Renaissance is one in which the rationalism of Enlightenment modernism is rejected for a reclamation of old world irrationalism. Mysticism replaces empiricism. This is a stance that is clearly and completely at odds with Einstein's cosmic religion, according to which the universe is an entirely well-ordered realm accessible to human reason. Einstein is the arch-empiricist. The two seem at loggerheads.

But Einstein, ever the peacemaker, finds common ground between his rationalism and Buber's irrationalism. Both start from the same shared experience, a sense of awe and wonder at the world and a desire to be a part of this something bigger than oneself.

> But the Jewish tradition also contains something else, something which finds splendid expression in many of the Psalms, namely, a sort of intoxicated joy and amazement at the beauty and grandeur of this world, of which man can form just a faint notion. This joy is the feeling from which true scientific research draws its spiritual sustenance, but which also seems to find expression in the songs of birds.[16]

The language that Einstein uses here would be readily recognized as the same sort of words that Buber uses in describing the Hasidic joy of being in the world.

> The core of hasidic teachings is the concept of a life of fervor, of exalted joy. But this teaching is not a theory which can persist regardless of whether it is translated into reality. It is rather the theoretic supplement to a life which was actually lived by the zaddikim and hasidim, especially in the first six generations of the movement.[17]

The routes they take from their common origin may be different, but assimilated and postassimilated Jews walk their distinct paths with a shared appreciation. In this union, Einstein argues, we find a commonality that we can begin to use to bridge the divide across Buber's dinner table that is played out fully in the divisions within the community at large.

Einstein's second foundational element of Judaism is the ethical posture one takes toward others.

> Judaism seems to me to be concerned almost exclusively with the moral attitude in life and to life. I look upon it as the essence of an attitude to life which is incarnate in the Jewish people rather than the essence of the laws laid down in the Torah and interpreted in the Talmud . . .
>
> The essence of that conception seems to me to lie in an affirmative attitude to the life of all creation. The life of the individual only has meaning insofar as it aids in making the life of every living thing nobler and more beautiful. Life is sacred, that is to say, it is the supreme value, to which all other values are subordinate. The hallowing of supra-individual life in its train a reverence for everything spiritual—a particularly characteristic feature of the Jewish tradition.
>
> . . . "Serving God" was equated with "serving the living." The best of the Jewish people, especially the Prophets and Jesus, contented tirelessly for this.[18]

This is Einstein's paraphrasing of Buber's ethical stance. Buber's Jewish existentialism is one in which the moral stance one must take in the world is not one of a legalistic adherence to Torah and Talmud but an understanding that we necessarily live in relationship with everyone else where we see the other person as a person and take their humanity as foremost.

> The human being to whom I say You I do not experience. But I stand in relation to him, in the sacred basic word . . .
>
> The You encounters me by grace—it cannot be found by seeking. But that I speak the basic word to it is a deed of my whole being, is my essential deed . . .
>
> The basic word I-You can be spoken only with one's whole being. The concentration and fusion into a whole being can never be accomplished by me, can never be accomplished without me. I require a You to become; becoming I, I say You.
>
> All actual life is encounter.[19]

While Einstein's thought had no effect on Buber, Buber clearly had an effect on Einstein.

A THIRD MOVEMENT ALSO ATTEMPTED to reconstruct a notion of Judaism, and that was one in which Judaism was neither a religion nor a community, but a nation. It is a modern usage of the word "nation" to imply nation-state, that is, a region with well-defined borders and a single government. The traditional concept of nation is broader, indicating a set of related communities spread out over a region. Consider, for example, the Nation of Islam or Red Sox Nation. Only after World War I, when the powers of Europe were rapidly decolonizing and creating the map of the world as we now see it, did nation and nation-state come to be thought of in the same way.

Around this time, those trying to define the notion of Judaism found themselves with more of a problem than they had initially considered. Secular Jews, of which there were many, were still to be considered Jews, so Judaism could not be reduced to a religion. These secular Jews had assimilated into society and physically relocated amidst non-Jewish neighbors, so Judaism was not defined simply by way of an appeal to a geographically coherent community. The lives of Eastern European Jewish immigrants escaping violent persecution had little in common with their Western middle- and upper-class counterparts whom they came to meet in their new homelands, so Judaism could not be considered a single culture. Yet, there was something that bound them all together. Judaism must be a nation.

And if all of these other nations were receiving their own states, why shouldn't the Jews? The pogroms of the east and the violence surrounding the Dreyfus affair in France made this politicization of Judaism seem ever more pressing. And so this time saw the rise of Zionism.

In 1897, Theodor Herzl organized the First Zionist Congress in Basel, Switzerland, a year after publishing *The Jewish State: An Attempt at a Modern Solution of the Jewish Question*. The idea was originally to populate Palestine with Jewish settlements as a safe haven, but the notion quickly transformed into the creation of a Jewish state. The modernist mind was oriented toward citizenship, and the idea of living somewhere necessarily brought with it questions of political power. One thing to which the Enlightenment thinkers seemed to give short shrift was the plight of the permanent minority. It was a problem the Jews of Europe knew all too well. And so as the globe was in the process of being divided up into new nation-states, surely Jews, the line went, deserved one more than any other nation as reparations for all of the suffering they have had to endure for generation after generation. As Ahad Ha'Am wrote:

> "According to the suffering, so is the reward." It cannot be that after thousands of years of untold evil and affliction the people of Israel will rejoice upon obtaining, at long last, to the rank of a small and mean nation, its state a plaything in the hands of great neighbours and incapable of survival except by the machinations of diplomacy and perpetual abasement before whomever fortune happens to have smiled upon; an ancient people which was a light unto gentiles cannot be satisfied with no more than this as a reward for its hardships—when many other nations, of unknown origins and without culture, have achieved it in short order without first suffering a fraction of what it had undergone.[20]

Ironically, Zionism brought Buber and Einstein together; ironic because while both emerged as significant spokesmen for the movement and both worked to organize and raise funds for Zionist causes, neither of them believed in the underlying claim that there should be a Jewish state. For both, Zionism was not a political position designed to secure control over land, but something completely different.

For Buber, Zionism was about the mystery at the core of Judaism:

> This is the theme, relating to a small and despised part of the human race and a small and desolate part of the earth, yet world-wide in its significance, that lies hidden in the name of Zion. It is not simply a special case among the national concepts and national movements: the exceptional quality that is here added to the universal makes it a unique cate-

gory extending far beyond the frontier of national problems and touching the domain of the universally human, the comic and of Being itself. In other respects, the people of Israel may be regarded as one of the many peoples on earth and the land of Israel as one land among other lands: but in their mutual relationship and in their common task they are unique and incomparable. And in spite of all the names and historical events that have come down to us, what has come to pass, what is coming and shall come to pass between them, is and remains a mystery. From generation to generation the Jewish people have never ceased to meditate on this mystery.[21]

For Einstein, in contrast, it was a much more pragmatic issue.[22] As long as Jews remained everywhere a minority, they would find security nowhere and risk either extermination or destruction by assimilation.

> Zionism . . . opens the prospect for a dignified human existence to many Jews who presently languish in Ukrainian hell or degenerate economically in Poland. By leading Jews back to Palestine and restoring a healthy and normal economic existence, Zionism represents a productive activity that enriches all of society. The main point, however, is that Zionism strengthens Jewish dignity and self-esteem, which are critical for existence in the Diaspora. Moreover, in establishing a Jewish center in Palestine it creates a strong bond that gives Jews a sense of self.[23]

While Buber and Einstein had different reasons for wanting to establish a lasting Jewish presence in Palestine, they both took issue with the political or nationalistic version of Zionism. Both Buber and Einstein were binationalists who opposed the Zionist vision of a Jewish state. The reason for both was that it ran contrary to the very ethos at the heart of Judaism. From Einstein:

> I should much rather see reasonable agreement with the Arabs on the basis of living together in peace than the creation of a Jewish state. Apart from practical considerations, my awareness of the essential nature of Judaism resists the idea of a Jewish state with borders, an army, and a

> measure of temporal power no matter how modest. I am afraid of the inner-damage Judaism will sustain—especially from the development of a narrow nationalism within our own ranks, against which we have already had to fight strongly, even without a Jewish state. We are no longer the Jews of the Maccabee period. A return to a nation in the political sense of the word would be the equivalent to turning away from the spiritualization which we owe to the genius of our prophets.[24]

Similarly, from Buber:

> Let us bear in mind—and actually there is no need for me to remind you, for our whole life is permeated by it—that other nations regarded us, and in some places still do, as alien and inferior. Let us beware of considering and behaving toward anyone who is foreign and as yet insufficiently known to us as if he were inferior! Let us beware of doing ourselves what has been done to us. Certainly, and again I emphasize this point, the maintenance of our existence is undoubtedly an essential prior condition of all our actions. But this is not enough. We also need imagination. Another thing we need is the ability to put ourselves in the place of the other individual, the stranger, and to make his soul ours.[25]

Indeed, Buber spent much time behind the scenes trying to undermine the rising idea of a Zionist defense force, a Jewish Legion, and the right-wing Revisionist group who took a militant stance against the Arabs in Palestine.[26]

So, while Einstein is readily and in a sense rightly identified as a Zionist, he also was alienated from the nationalism at the heart of the movement. Again, despite this identification, we find no sense in which Einstein's thought is influencing the Jewish debate here.

WHILE THE REST OF THE WORLD seemed to think that Einstein's theory of relativity changed everything, for those Jewish scholars engaged in the various elements of the primary questions concerning the explicitly Jewish thinkers, it seemed to change nothing. Einstein was not relevant at all to those who were actively engaged in the deep questions that arose from the Jewish intellectual context of the time.

The opposite, however, was the case, with those thinkers of Jewish heritage who were concerned with the questions that seemed most important outside of the Jewish community. To assimilated Jewish intellectuals, Einstein was crucial.

Hans Reichenbach was one of eight students to take Einstein's first seminar in the general theory of relativity at the University of Berlin in 1919.[27] He was trained as an engineer and worked in the army signal corps during the world war, then as an engineer working on radio technology until he began teaching engineering. But his interest turned to philosophy.

Reichenbach was among the first philosophers to realize exactly how radical the theory of relativity was and how deep its philosophical ramifications ran. Philosophers of that period who had any interest in science based their views on the work of Kant, who believed that the roots of all scientific understanding of the world were hardwired into the human mind. Kant argued that observations do not just appear in the mind like a digital picture on a screen. The senses give the raw matter of observation to the mind, but the mind has to shape it, massage it into form, in order to have a well-developed observation. Seeing something requires that the eyes give the mind an undifferentiated blur of colors that the mind then takes and turns into individuated objects situated in certain geometrical relations. When I hold my thumb up by the moon, I know that my thumb is closer because it is bigger or because it obscures part of the moon. But how do I know that the moon isn't closer, but really, really small? Or that the moon hasn't become part of my thumb? We just know. This can't be from the senses because both are interpretations of the same vision; it must be because the mind plays an active part in constructing what we observe.

According to Kant, the instruction manual for how to go about constructing these observations includes Newtonian physics. It's not that it Newton's laws just happen to be true of the world; rather, they are the rules by which our minds create the world we see. They are preconditions to the possibility of observing anything. And as such, they are undeniably true. No observation could ever undermine them because they are the instructions by which all observations are created. No human mind could ever possibly think of the world any differently. It is part of being human that we have to see it that way.

And yet here was Einstein saying that Newton's laws were wrong and supporting the claim with empirical evidence. On top of this, he was also using different basic concepts, ideas Kant held that no human mind could entertain. If Kant's theoretical psychology was correct, Einstein would not only be wrong, his theory would be impossible for a human to conceive. It shouldn't be able to happen. But there it was. And lying in smoldering ruin beside it was the accepted philosophical view of science. And no one seemed to realize it.

So Reichenbach would have to tell them. He would have to figure out how to reform what we thought were the basic concepts by which we make sense of the world and the basic sorts of inferences we use. But in reconstructing our philosophical understanding of the roots of science, he needed to avoid the problem Kant walked into—tying his philosophical system to a particular scientific theory. Einstein's theory was wonderful, but the day would come when it, too, would be overturned. Science progresses, it grows, and any philosophical account worth considering will have to be able to make sense of this progress and not get derailed by it.

Reichenbach found this by combining a strict axiomatic approach that lays out clearly all presuppositions with the style of reasoning we find in Einstein. Think back to Einstein's argument about the magnet and coil at the beginning of "On the Electrodynamics of Moving Bodies." Where Maxwell gave different descriptions of reality based on whether the magnet or the coil was *really* moving, Einstein said that since you observe the same thing either way, there is no "really" question here, that the two accounts must be saying the same thing about nature, just saying it in different ways. Reichenbach adopts the same sort of view at the next higher level of abstraction. Where Einstein considers the meanings of measurements within a theory, Reichenbach looks at the meanings of the theories themselves. If we consider two theories that make all the same predictions, but do it with different sets of concepts, then, Reichenbach argues, philosophically, the two theories must really be different versions of same theory, just expressed in different languages.

What Einstein argues within a theory, Reichenbach argues about theories.

Reichenbach sees the job of the philosopher as being the house maid of the scientist. The scientist creates a theory in a way that is intuitive and sloppy. If scientists had to fret about exactitude and details, they would not be able to summon the creativity needed for great advances. Once a hypothesis is set out, evidence is gathered to support it. If enough evidence is found and scientists start taking the theory seriously, the philosopher then has an assignment. The philosopher, Reichenbach argues, comes in after the fact and tidies it all up, replacing the loose notions and slapdash arguments of the scientists with clearly defined concepts and exacting axioms, transforming the theory into a tight, rigorous, logical machine in which the basic observable facts of the world are set down in one column and the theoretical, conceptual elements are set down in the other with absolute care for clarity. In this way, philosophers rationally reconstruct scientific theories into sets of axiomatic propositions that are then used to construct a possible world.

Kant was right, Reichenbach contends, that we need an intellectual addition to the raw input of our senses to create observations; but Kant was wrong to think there can be but one way of doing it.[30] Every scientific theory is a new way of creating a world. Some create worlds that are closer to our world; others create worlds that are radically different. And if two theories create the same set of possible worlds, then, even if their inner workings seem different, we are to consider them as different ways of saying the same thing about the world. And by considering these tidied up, axiomatic versions of the theories, we can see exactly where two different theories say different things and how to develop a test that would determine which should be kept and which rejected.[31]

Reichenbach spent two years coming up with this sort of axiomatic construction of the special theory of relativity[32] and used it to defend Einstein against his critics of all sorts in the philosophy journals, the physics journals, and the popular press.[33] He would become one of Einstein's bulldogs and a successful popularizer, publishing articles in science magazines, writing books for the nonscientist, and hosting his own radio program explaining modern science.

When Reichenbach wanted to move from the technical university in Stuttgart, where he had to split his time teaching philosophy of science and engineering courses, to Berlin where he would be solely a philosopher, the philosophy faculty rejected him arguing that his ideas were not philosophy as they understood it. Einstein himself went to them and argued that their conservative view was stifling the field they loved. But the argument failed to sway them. So Einstein did an end run around them and convinced Planck that the physics department needed a chair in the foundations of physics, a chair to be occupied by a philosopher with a significant background in modern science. And so Reichenbach came to Berlin.

He stayed there until 1933, when Reichenbach, a *Mischling*, was a victim of the Nazi purge of the universities and fled for several years to Turkey. There Mustafa Kamal Attaturk cleverly realized that the rejection of so many German intellectuals would allow him to bring in world-class talent on the cheap to train his own people. German thinkers from across the academic spectrum were brought in to teach the next generation of Turkish university professors. A safe place and a few translators was all that were needed, and Attaturk made sure they were both in plentiful supply.

It was a fortunate break for Reichenbach, who foresaw the purge and had secured his place in the Turkish system just in case. But others were not so lucky. Reichenbach's student, Kurt Grelling, a promising logician and philosopher of science, was captured by the Nazis and held in a concentration camp in France until he was shipped to Auschwitz along with his wife, a non-Jew who refused to abandon her beloved husband.[34]

Moritz Schlick was a colleague of Reichenbach's heading up the group of philosophers of science in Vienna. Schlick, whose Ph.D. work in physics was done under the watchful eye of Planck, wrote the first philosophical book discussing the details of relativity theory, *Space and Time in Modern Physics*, a book Einstein personally congratulated him on. Einstein was so impressed that the two began corresponding, and Einstein stayed with the Schlicks when visiting Vienna.[35]

Schlick was assassinated, shot to death at the university by a former student of his, Johan Nelböck. In the press, the murder was lauded by Austrian fascists as a strike against the rising "Jewish philosophy." In the Catholic weekly magazine, *Schönere Zukunft*, Johannes Sauter, a professor of law and sociology, published an extended article "The Case of Professor Schlick in Vienna—A Reminder to Search Our Conscience" under the pseudonym, "Dr. Austriacus."

> It is well-known that Schlick, whose research assistants were a Jewish man and two Jewish women, was the idol of Vienna's Jewish circles. And now the Jewish circles of Vienna are constantly celebrating him as the most significant thinker. This we understand very well. For the Jew is the born anti-metaphysician and loves Logicism, Mathematicism, Formalism, and Positivism in philosophy—all of them qualities which Schlick possessed in abundance. Nevertheless we would like to call to mind that we are Christians living in a Christian-German state, and that it is we who will decide which philosophy is good and appropriate. Let the Jews have their Jewish philosophers at their Cultural Institute! But the philosophical chairs at the University of Vienna in Christian-German Austria should be held by Christian philosophers! It has been declared on numerous occasions recently that a peaceful solution to the Jewish question in Austria is also in the interest of the Jews themselves, since a violent solution of that question would be unavoidable otherwise. It is to be hoped that the terrible murder at the University of Vienna will quicken efforts to find a truly satisfactory solution of the Jewish question![36]

Schlick was not Jewish. He was not a political rabble rouser. Like his mentor Planck, he was a gentleman in the traditional Prussian mold, respectable and respectful. But the philosophy that derived from Einstein's "Jewish science" became seen by the same people as "Jewish philosophy" and that led to tragic consequences.

The irony though is while this philosophical movement derived directly from the thought of Einstein, he rejected it.[37] When Philipp Frank, one of Einstein's friends and a leader of the new philosophical movement, asked him how he could reject an intellectual revolution he had started, Einstein replied that "a good joke should not be repeated too often."[38] Indeed, it was a joke the German nationalists failed to find any humor in and even Einstein's personal disavowal would do nothing for the sad consequences these thinkers would suffer.

BUT ONE DID NOT NEED TO BE AN INTELLECTUAL heir to Einstein to be subject to oppression. This was a lesson sadly taught to Edmund Husserl, whose philosophical project, phenomenology, is the foundation of existentialism, perhaps the most well-known philosophical movement of the twentieth century. Husserl was a mathematician by training, who made the move later in life to philosophy. He was, for a while, a colleague of David Hilbert's at the University of Göttingen, where their children were classmates. They clashed with each other over intellectual and university-related issues. Ultimately, Husserl moved to the University of Freiburg where he retired and was a cherished emeritus professor until this was stripped from him in 1933.

Husserl, a Jew from the Jewish enclave of Prossnitz, was entirely assimilated and had had himself baptized. Although he never would have self-identified as Jewish, when it was time to marry, he returned to Prossnitz and found his future wife Malvine Steinschneider, who was also Jewish, although she too was baptized soon after the wedding. The Husserls did not think of themselves as Jewish at all. They were Germans. In the time of war, all of their three children joined the effort, sons Wolfgang and Gerhart as front-line soldiers and daughter Elizabeth as a nurse. Only Gerhart and Elizabeth came home. Wolfgang died at Verdun, a loss that devastated Husserl.

Husserl's philosophical train of thought starts with a problem that Descartes wrestled with—the problem of the evil demon. Recall that Descartes sought absolute certainty, that which we can know beyond any doubt. He considered the possibility that his mind was controlled by an evil demon who fed in every thought, every idea, every sensation, none of it connecting with the actual external world. Given that I only have access to the set of internal ideas of my mind, how can I know that any of them reflect the actual state of reality? Philosophers, following Kant, use the word "phenomena" to refer to the internal experience of a thing and "noumena" when talking about the thing itself. Descartes is asking whether our experience of the phenomena allows us to infer absolutely anything at all about the noumena.

Descartes seems to be trapped. It seems that there is no way to know anything with absolute certainty, until Descartes realizes that there is something that even the evil demon could not fool him about—that he exists, because he would have to exist as a thing that thinks in order to be fooled. So, Descartes argues, the act of doubting entails with absolute certainty that I exist. I think therefore I am, *cogito ergo sum.*

Descartes then goes on to try to restore virtually all other beliefs by proving the existence of an all-perfect, and thereby non-deceiving, God. This argument was known for centuries to be deeply flawed. Husserl rejects everything beyond this move but holds that Descartes's *cogito* is right on the mark. The only thing we can know with absolute certainty is that we exist as things that think. We know that we think and we know what we think, but any attempt to extrapolate the content of our thoughts beyond our own experience is deeply problematic. We can only speak in terms of the phenomena. We live a life and have experiences and, when we refer to things, we make reference only to those objects we encounter, not some set of artificial, abstract entities.

> the world, or rather individual things in the world as absolute, are replaced by the respective meaning of each in *consciousness* in its various modes (perceptual meaning, recollected meaning, and so on).[39]

All of human thought is expressing, analyzing, comparing, categorizing, and understanding our internal experiences. Human phenomena are the sole building blocks of all we know, of all we can know.

But when shopping for a car, I walk around it. I sit inside it. I kick the tires. Each of these provides me with different experiences, yet I believe that they are all aspects of what I consider to be a single thing. That thing is not posited to be an external object, but rather constructed from my experiences. We hold certain elements from distinct experiences to be aspects of a form that makes up the content of the various experiences. Husserl uses the word "eidetic" to refer to these unchanging elements of our varied experiences. What science ought to do, Husserl argues, is take these eidetic building blocks and study them, categorize them, find relations between them. Science is to start with and account for our experiences. Science is a human-based endeavor that makes sense of human phenomena for the sake of human living.

Beneath the eidetic elements, Husserl argues, still more basic preconditions underly the possibility of experience. These "transcendental" elements, things like time and space, are implicitly a part of all human experiences; we cannot experience anything timelessly or outside of any notion of place. The eidetic elements are present in multiple experiences, but the transcendental ones must be implicit in all human experiences.

Modern science, Husserl contends, has gone completely awry in two crucial ways. First, it seeks objective truth, trying to find and organize facts about an external world. It seeks to do it in a way that does not require any reference to the subjective experience of a human observer. Second, in removing scientific results from the "life-world," it completely misunderstands what it does and does not do. Its interpretation of its own work becomes radically flawed. Husserl argues that there is a crisis in European science[40]—the crisis that led to the dehumanizing tragedy that was the Great War—because people are starting to believe that they have knowledge, both physical and metaphysical, about some external reality based on modern science in a fashion that removes scientific knowledge from the life-world.

The prime example of someone making both of these mistakes? Albert Einstein. We see in Einstein, he contends, a false move away from human experience as the basis of science.

> The sciences build upon the life-world as taken for granted in that they make use of whatever in it happens to be necessary for their particular ends. But to use the life-world in this way is not to know it scientifically in its own manner of being. For example, Einstein uses the Michelson experiments . . . There is no doubt that everything that enters in here—the persons, the apparatus, the room in the institute, etc.—can itself become a subject of investigation in the usual sense of objective inquiry, that of the positive sciences. But Einstein could make no use whatever of a theoretical psychological-psychophysical construction of the objective being of Mr. Michelson; rather, he made use of the human being who was accessible to him, as to everyone else in the prescientific world, as an object of straightforward experience, the human being whose existence, with this vitality, in these activities and creations within the common life-world, is always the presupposition for all of Einstein's objective-scientific lines of inquiry, projects, and accomplishments pertaining to Michelson's experiments.[41]

Einstein himself contends that his work in physics is a move away from the "merely-personal" to something universal and objective. Husserl contends that no such escape is possible.

> Precisely this world and everything that happens in it, used as needed for scientific and other ends, bears, on the other hand, for every natural scientist in his thematic orientation towards its "objective truth," the stamp "merely subjective and relative." The contrast to this determines, as we said, the sense of the "objective" task. This "subjective-relative" is supposed to be "overcome"; one can and should correlate with it a hypothetical being-in-itself, a substrate for logical-mathematical "truths-in-themselves" that one can approximate through even newer and better hypothetical approaches, always justifying them through experimental verification. This is the one side. But while the natural scientist is thus

> interested in the objective and is involved in this activity, the subjective-relative is on the other hand still functioning for him, not as something irrelevant that must be passed through but as that which ultimately grounds the theoretical-logical ontic validity for all objective verification, i.e., as the source of self-evidence, the source of verification. The visible measuring scales, scale-markings, etc., are used as actually existing things, not as illusions; thus that which actually exists in the life-world, as something valid, is a premise.[42]

This is no benign error by Einstein. By pretending he is dealing with objectivity by ignoring the essential elements of the life-world, Einstein fails to understand what his theory does and does not mean.

> Einstein's revolutionary innovations concern the formulae through which the idealized and naively objectified *physics* is dealt with. But how formulae in general, how mathematical objectification in general, receive meaning on the foundation of life and the intuitively given surrounding world—of this we learn nothing; and thus Einstein does not reform the space and time in which our vital life runs its course.
>
> Mathematical natural science is a wonderful technique for making inductions with an efficiency, a degree of probability, a precision, and a computability that were simply unimaginable in earlier times. As an accomplishment it is a triumph of the human spirit. As for the rationality of its methods and theories, however, it is a thoroughly relative one. It even presupposes a fundamental approach that is itself totally lacking in rationality.[43]

Descartes tried unsuccessfully to move beyond human experience to the world itself, from the phenomena to the noumena, by giving an unsound proof for the existence of the Christian God. Einstein, Husserl argues, makes exactly the same error by substituting a scientific method in the place of the Cartesian theology. Both, he claims, wrongly try to move outside the mind to a false objectivity instead of embracing the relative-subjective, the phenomenological, the fact that the life-world *is* the world. To lose sight of that is one step on a slippery slope to barbarism.

Martin Heidegger, Husserl's star student thought that even Husserl's narrowed focus was too broad. The real question of philosophy, he asserted, is not what it is for us to be the beings we think we are, people who have a hectic life full of things we can and cannot control. Rather the real concern of philosophy is the ultimate fundamental question, what he calls the "ontological question," "What is Being?"

As people, we are thrown into a world we did not choose, into historical and social contexts beyond our control, to live a life we did not ask for. Humans are essentially temporal beings, both because we live in our times, but also because we live as future-directed beings. We have plans and projects. We are the people we picture ourselves as becoming. This means that when we look at the world around us, we make sense of it in terms of where we think we are going and how the things in it can be useful to us in getting there. Unlike Husserl, Heidegger argues that our experience of things is always mediated through the way we use things and what we use them for; we cannot remove these and find our way to some eidetic ideal lying beneath.

But this is being (with a mere lowercase b) in everyday life where we elevate the mundane and meaningless to a status they do not deserve. The true question of philosophy is more basic. What is Being (with a much more impressive capital B) itself? As beings we are pulled in many different directions, responding to the expectations and needs of the masses, and all of this distracts us, takes us away from being able to see our own Being. The understanding of this is the single task of philosophy and the possibility of engaging in this philosophical undertaking is what makes us what Heidegger calls *Dasein*—Being-there.

> *Dasein* is an entity which does not just occur among other entities. Rather, it is ontically distinguished by the fact that, in its very Being, that Being is an issue for it. But in that case, this is a constitutive state of *Dasein*'s Being, and this implies that *Dasein*, in its Being, has a relationship towards that Being—a relationship which in itself is one of Being. And this means further that there is some way in which *Dasein*

> understands itself in its Being, and that to some degree it does so explicitly. It is peculiar to this entity that with and through its Being, this Being is disclosed to it. *Understanding of Being is itself a definite characteristic of* Dasein's *Being.*[44]

The noise of being irreparably obscures any sense of Being. To do the deep, difficult, philosophical work we are driven to do, we need to occupy a specific perspective. We need to completely disrupt our normal way of seeing and the only thing that will do this for us is the moment when we grasp our own mortality. When we see that our time is limited, we no longer see the self that we project into the future in terms of our plans and projects. When we see that there is a moment when we cease to be, at that point alone can we look from a perspective in which we can grasp ourselves as Being through time and not as a mere being in time. It is only then that we are no longer the temporally fragmented being that occupies various positions in time and can be viewed by the self as the complete unified Being we are. In the angst and dread of the moment when we truly make sense of our limitedness in time, we are finally able to free ourselves from the banalities of daily life that populate our mind and come to consider Being.

Modern science and the technology it generates are useful in the lives of beings but bring with them, Heidegger argues, a stance that is inimical to the philosophical task of understanding Being. Newton mathematizes time, makes it a pure concept through which one can relate the behaviors of point masses. This conceptual time is not the time of Being and making the metaphysical claim that it is the true time steers us away from the standpoint from which philosophy could actually be done.

> The threat to man does not come in the first instance from the potentially lethal machines and apparatus of technology. The actual threat has already afflicted man in his essence. The rule of enframing threatens man with the possibility that it could be denied to him to enter into a more original revealing and hence to experience the call of a more primal truth.[45]

Science is important to society, Heidegger admits, but in scientizing the universe, it makes the essential task of philosophizing impossible. "Scientific research is not the only manner of Being which this entity can have, nor is it the one which lies the closest."[46]

Heidegger was Husserl's assistant for years, and Husserl came to view him as part of the family. He championed Heidegger's work and career, something that was essential for getting a position in the government-controlled bureaucratic German university system. Husserl spent a lot of his political capital on Heidegger, ultimately bringing him in as the successor to his own prestigious chair at the University of Freiburg. From this position, Heidegger was elevated to the position of rector, and then, to the astonishment of Husserl, became a member of the Nazi Party.

Many myths surround the events between Heidegger and Husserl and radically different accounts of what happened exist, some from Heidegger himself after the fact and others from historians. What we do know is that Heidegger became rector of the university. We know that Husserl received notice that he was no longer associated with the university, his rights and privileges as an emeritus professor having been revoked. We know that Heidegger became a member of the Nazi Party while rector. We know that Heidegger removed the dedication to Husserl from his masterwork *Being and Time*. We know that Heidegger did not attend Husserl's funeral and that there was as a result no further relationship between the families. But much else that is widely believed is false or contested.[47]

The widespread myth is that Heidegger had Husserl banned from campus or the university library. This is false. But contrary to his protestations years later, this in no way means that Heidegger was innocent.

At the beginning of April of 1933, the faculty of Freiburg elected Wilhelm von Möllendorf, a professor of anatomy, as rector of the university. Less than a week later, the local authorities announced a purge of non-Aryan civil servants, including university professors. This meant that Husserl, who was retired, would lose his place as a part of the institution. In protest, von Möllendorf resigned.

In later accounts, Heidegger claimed that von Möllendorf, who lived near the Heideggers, and Josef Sauer, the previous rector, came to him and asked him to step forward as the next rector, something he was reticent to do given his lack of administrative experience.[48] Historian Hugo Ott, however, produces evidence that prior to von Möllendorf's resignation, a group of faculty members who were either members of the Nazi Party or sympathizers had already begun behind the scenes to plot to undermine von Möllendorf, a Social Democrat.[49] They decided among themselves that someone from their own ranks, someone with strong Nazi leanings, should be the next rector in order to help transform the institution in a way that suited their Hitler-inspired worldview. That man, they decided, should be Heidegger.

So, when von Möllendorf resigned, this group sprung into action and given the executive order, all Jewish faculty members lost voting privileges, allowing them to ram Heidegger into office at an unprecedented pace with minimal discussion. And so Heidegger assumed command of the university. His installation brought out Nazi dignitaries and brown shirts were worn openly at the ceremony. Heidegger's inaugural address was entitled, "The Self-Assertion of the German Universities," in which he argues that there are three services to which the academic community must be committed for the sake of the Reich—labor service, army service, and knowledge service:

> The German Universities will only attain power if the three forms of service unite in overwhelming force, if both teachers and students in their adherence to tradition place themselves side by side in the thick of the fight. All the powers of the heart and all the skills of the body are developed through struggle, are strengthened in battle, and are proven by continuing the fight. *We first understand the glory and the greatness of the Hitler revolution* when we carry implanted deep within us this reflection: Everything that is great is in the midst of the storm.[50]

One might think, despite Heidegger's National Socialist leanings, that he would object to the ill-treatment of his own mentor and patron. After all, Heidegger did formally protest in the cases of two other Jewish faculty members—one was an important scholar whose loss would damage the university's reputation and the other a physical chemist who had a grant that paid for all of the equipment at the university's Institute for Physical Chemistry and his loss would gut the university's chemistry program.[51] But when it came to Husserl, the man who taught him and fought to get him his job, it was not merely silence, but worse.

The only communication between the two was a letter sent from Heidegger's wife to Husserl's wife in which Husserl's removal from the university is never mentioned, only the firing of their son Gerhart who had become a professor of law at a different university that receives a mere casual and cold mention.[52] There was no offer of support for the man who had done so much for them. There was no expression of outrage or sympathy for his predicament. There is no acknowledgement at all about his treatment with respect to the institution where he had spent so many years working and helping Heidegger become what he became, the rector of that very institution. Indeed, the tone of the letter is peculiarly formal for families who had been so close. The formality was a clear signal that the personal relationship was being severed. After all Husserl had done for him, Heidegger had no interesting in maintaining contact with his mentor, because it had become inconvenient to Heidegger's professional progress that Husserl was Jewish.

The irony here is that the local purge of non-Aryan civil servants that caused Husserl to be removed from his place in the university was ultimately nullified by the *Reichsgesetzblatt*, the national law purging non-Aryan civil servants. Recall that Hitler's version had run into the objections of von Hindenberg and was altered to allow exceptions for those who served at the front in the Great War, or who lost a father or son in battle. Heidegger, the brash supporter of the rise of German power, served minimally in the war as a home-stationed censor of mail because he was physically unfit for any real service. Meanwhile, the Husserl family had sent all of its children into battle and made the ultimate sacrifice of one of its sons, which in 1933 qualified Husserl to maintain his rights and privileges at the university. This is something that Heidegger would have known as rector and family friend but never bothered to mention to him.

While Husserl was no longer officially persona non grata at Freiburg, Heidegger's wife's letter was sent just days before Heidegger was to ostentatiously step forward and join the Nazi Party in a garish ceremony on International Worker's Day, May 1, something that was planned well in advance. Once in power and a member of the party, Heidegger decided to use the university to hold a summer camp to train student leaders in the National Socialist ways[53] and rewrote the school constitution, overturning the democratic model of faculty rule in exactly the same way that the Weimar Republic had been replaced. No longer was there to be an elected rector with a fixed term of service, now there would be a Führer-rector with complete control over appointment of deans.[54] Heidegger, of course, became the Führer of the University of Freiburg and in that role participated with the Gestapo in having a member of the faculty investigated for pacifist leanings. Indeed, his henchman in the investigations was a physicist who had been a student of Phillip Lenard, the father of the Aryan physics movement.[55]

Husserl may have been legally entitled to return to the university, but Heidegger made sure that it would be a place to which he would not be welcome. And when Husserl died after a protracted illness, Heidegger neither attended the funeral nor sent any condolences. Husserl was no longer needed, he had served his purpose for Heidegger, he was no longer ready-to-hand. And so even those who opposed Einstein's views were not safe from Nazi persecution.

THUS WE CAN SAY THAT WHEN WE TAKE "Jewish science" to be a scientific element of explicitly Jewish thought of Einstein's time, the theory of relativity is clearly not a development of interest to those whose concern was to maintain a clearly and explicitly Jewish intellectual movement within the Jewish community. Some found it harmful to the community—a force for modernism and secularization—others simply saw it as irrelevant. But when we look at secular, assimilated Jews, here we find deep thinkers wrestling on all sides of Einstein's work. On one hand, you have those irreligious

Jews who championed Einstein and turned his work into a foundation for philosophical thought. They were persecuted by the Nazis. On the other hand, you have thinkers of Jewish background who rejected Einstein as a foundation for philosophical thought. They were also persecuted by the Nazis.

CHAPTER SIX

Einstein's Liberal Science?

It has been more than a century since Einstein's theory was introduced in "On the Electrodynamics of Moving Bodies." The cutting edge of physics is well beyond relativity. The Nazis are now relegated to History Channel theme weeks, sight gags in Mel Brooks films, and ad hominem attacks on talk radio. Surely, the argument about Einstein's theory being Jewish science is all just an interesting historical episode by now.

Actually, there is still opposition to the theory of relativity from a wing of the contemporary American conservative movement based in some cases on its association with Albert Einstein and the perception that it is Jewish science, or to change with the times, "liberal science." It certainly is not the case that these contemporary opponents of relativity adhere to the beliefs of National Socialism or are in any fashion sympathetic toward its tenets, but it is interesting that some of their arguments are updated versions of those found in Weyland, Stark, and Lenard.

We find a modern version of "One Hundred Authors against Einstein" in the entries for "Theory of Relativity"[1] and "Counterexamples to Relativity"[2] on the conservative wiki site, Conservapedia, designed as an alternative to Wikipedia in which articles are written from a conservative point of view. Its treatment of the theory of relativity is an extended anti-Einstein and antirelativity polemic reminiscent of the wide range of voices and approaches marshaled against the theory in the 1920s and 1930s in which any argument against relativity was thrown onto the stage. We find four themes running throughout the treatment: (1) Einstein is not deserving of any credit for its discovery or successes, should there be any; (2) the theory is not about the world but about formal symbols; (3) the theory is unsupported by empirical data; and (4) competing theories are suppressed because of a concerted political effort on the part of pro-relativity supporters. All are reminiscent of arguments seen between the world wars.

The article on the theory of relativity begins by distinguishing the special from the general theory and in both cases, making sure to list Einstein last among those deserving credit for developing the theory. Just as we saw with Lenard's book of great scientists, relativity would be rejected, but if there is any glory to be had from it care is taken to place it with someone other than Einstein. Indeed, in the section explaining the special theory of relativity, there is but one mention of the name "Einstein" anywhere in the text and that is to attribute the theory to someone else.

> Lorentz and Poincaré developed Special Relativity as a way of understanding how Maxwell's equations for electromagnetism could be valid in different frames of reference. Einstein famously published an explanation of Poincaré's theory in terms of two assumptions (postulates).[3]

It is, on this account, Lorentz's and Poincaré's theory, not Einstein's.

This is an odd claim in that not only did neither Lorentz nor Poincaré seek priority in the discovery but rather both were openly hostile toward it, something Einstein worked hard to change. Einstein tried on several occasions to convince the two. While his efforts with Poincaré repeatedly failed,[4] things went better with Lorentz. He gave his famed address "Ether and the Theory of Relativity" in 1920 at the University of Leyden knowing that Lorentz would be in the audience. One cannot but read the talk as anything but an attempt to show Lorentz the debt Einstein owed him and the ways in which one could make the theory philosophically consistent with Lorentz's less revolutionary bearing. It worked. Lorentz later "unequivocally emphasized Einstein's originality"[5] in his own book *The Einstein Theory of Relativity: A Concise Statement*:

> It is comprehensible that a person could not have arrived at such a far-reaching change of view by continuing to follow the old beaten paths, but only by introducing some sort of new idea. Indeed, Einstein arrived at his theory through a train of thought of great originality.[6]

One could scarcely ask for words that sound less like those of someone in a priority dispute over a theory.

The anti-Einstein bias in the piece is clear. The article is twelve pages long, and Einstein's name only appears twelve times. Of those occurrences, two mentions appear in explaining the elevator thought experiment connected with the principle of equivalence in the general theory of relativity, but the other ten all occur in contexts that are negative toward Einstein. Several times Einstein is only mentioned to point out those who reject his theory or its application, such as here:

> In recent years most physicists have shifted away from Einstein's original reliance on relativistic mass and his suggestion that mass increases.[7]

And here:

> Lorentz has this to say on the discrepancies between the empirical eclipse data and Einstein's predictions. "It indeed seems that the discrepancies may be ascribed to faults in observations, which supposition is supported by the fact that the observations at Prince's Island, which, it is true, did not turn out quite as well as those mentioned above, gave the result, of 1.64, somewhat lower than Einstein's figure."[8]

And here:

> Tom Van Flandern, an astronomer hired to work on GPS in the late 1990s, concluded that "the GPS programmers don't need relativity." He was quoted as saying that the GPS programmers "have basically blown off Einstein."[9]

Others are to try to undermine confidence in the theory:

> A dramatic but later discredited claim by Sir Arthur Eddington of experimental proof of General Relativity in 1919 made Einstein a household name.[10]

And still others to direct credit elsewhere:

> Other engineers and scientists have written about problems in the basic set of special relativity equations. Based on the ideas of not Einstein but of the scientist Fitzgerald as well as others, a length contraction effect was predicted as an explanation of the failure of the Michelson Morley experiment.[11]

Clearly, the theory most associated with Einstein's name should not, in fact, lead to his being celebrated at all.

The second thread of antirelativistic argumentation is a revival of the old antimodernist argument against the theory, that is, that the theory of relativity is not about nature, but about formalism. The theory takes us away from the world itself into some odd mathematical puzzle masquerading as reality.

> More generally, and also unlike most of physics, the theories of relativity consist of complex mathematical equations relying on several hypotheses. For example, at Hofstra University general relativity is taught as part of an upperclass math course on differential geometry, based on three stated assumptions.[12]

The topic's being dealt with in a mathematics course is taken as evidence that it is not truly a theory about the universe, just about the formalism. Of course, many mathematics classes deal with applications of the mathematics they consider. Psychology majors have to take statistical methods courses. Engineers must take calculus courses. The fact that mathematicians teach a course dealing with the sort of equations one finds in a field does not mean that the calculations studied have no application in the real world.

Further, the theory of relativity is not at all "unlike most of physics" in having a set of foundational complex mathematical equations relying on hypotheses. Indeed, every major physical theory since Newtonian mechanics has had such a form. It is true that the theory of general relativity, with its use of tensors requires mathematics that earlier theories did not. But Newton's theory requires calculus, mathematical machinery that earlier theories did not. Quantum mechanics, which is embraced throughout the treatment for its conflicts with relativity theory, also has a number of novel mathematical tools in its toolbox such as the Dirac delta function, which is an infinitesimally thin, infinitely tall curve, with an area of 1 underneath it.

Third, a significant amount of space is dedicated to mentioning results that seem to undermine the theory of relativity's empirical success. Examples are piled up citing cases in which the theory's predictions about observable phenomena do not exactly match measurement. For example, several disparaging mentions are made of Arthur Eddington's eclipse observations, claiming them to have been "discredited" because Eddington massaged the data to make sure Einstein's theory prevailed.

This is a shifty move on two fronts. First, it is not clear that it is true. John Earman and Clark Glymour, prominent scholars in the history and philosophy of physics, argue that Eddington and Sir Frank Watson Dyson played fast and loose with the data they received from the 1919 eclipse experiments.[13] Because of clouds and telescopic problems, they had to interpret and then count and discount data sets and did so in a way that seems to have favored Einstein. This, they argue, should have made their proclamation weaker than they actually declared it to be, that is, as an unequivocal vindication of Einstein. Others, like physicist Daniel Kennefick, argue that the inferences made were perfectly rational and normal means of data reduction.[14] So, at best the case is ambiguous.

But that ambiguity only remains if we limit ourselves to the 1919 observations. If, as is much more subtly mentioned just once in the Conservapedia article, we consider other similar observations made of light bending during later eclipses using better and more accurate equipment, then we would find evidence for relativity theory, not against it. The measurements that support the theory are buried, while more provocative claims are left intact and placed in prominent positions in the text such that an unknowing reader could easily be led to the wrong conclusion. This could be attributed to the wiki's open editing. Multiple authors working on the same text is a situation that is bound to have unfortunate consequences like this. Or it could be intentional rhetorical trickery. I don't claim either way, but it is something that should strike the critical reader as odd.

Some objections give no actual grounds for rejecting the theory of relativity but rely instead solely on arguments from authority:

> Louis Essen, the man credited with determining the speed of light, wrote many fiery papers against [relativity theory] such as *The Special Theory of Relativity: A Critical Analysis*.[15]

No sense of Essen's objections is considered and no discussion is given concerning why most physicists reject them. Similarly, we find a reference to the work of theoretician Robert Dicke:

> Physicist Robert Dicke of Princeton University was a prominent critic of general relativity, and Dicke's alternative "has enjoyed a renaissance in connection with theories of higher dimensional space-time."[16]

What the author strangely omits is that the quotation about Dicke's competing theory comes from an article in *Time Magazine* entitled "A Victory for Relativity"[17] which explains how observations from the Mariner space probes vindicated Einstein's theory over Dicke's.

Several other seemingly empirical-based objections curiously admit that relativity accounts for the data but argue that other competing theories come close to doing it as well. General relativity predicts the advance of the perihelion of Mercury, or its closest orbit to the sun. While "Newton's theory (assuming a precise inverse-square relationship for distance) predicts a rate of precession that differs from the measured rate by approximately 43 arcseconds per century,"[18] the authors claim that it is still the case that "Newton's theory can also explain this perihelion by factoring in the gravitational pull due to other planets."[19]

Similarly, while the general theory of relativity gives us successful predictions about gravitational lensing, we should, "note, however, that the extent of bending of light predicted by Newton's theory is open to debate, and depends on assumptions about the nature of light for gravitational purposes."[20] Newton's theory is close and since my team came in second, you should not award the trophy to the other team just because it had so many more points in the championship game.

One element of relativity is looked on favorably: time dilation. That time intervals lengthen from moving reference frames is seen as desirable:

> Creation scientists such as physicists Dr. Russell Humphreys and Dr. John Harnett have used relativistic time dilation to explain how the earth can be only 6,000 years old even though cosmological data (background radiation, supernovae, etc.) set a much older age for the universe.[21]

Similarly, the "action-at-a-distance by Jesus, described in John 4:46–54" is held up as a counterexample to relativity, that "shows that the theory is incorrect."[22]

The fourth line of attack on relativity is a version of Stark's argument that the widespread unwarranted support for relativity theory is the result of political control by pro-Einstein advocates. "There is," they contend, "an unmistakable effort to censor or ostracize criticism of relativity."[23] We see from the Conservapedia articles that there remain certain strains of anti-Einstein and antirelativity animus still operative among some conservatives.

WE FIND EVEN MORE VOCIFEROUS anti-Einstein arguments with the writings of Christopher Jon Bjerknes. He is the author of the blog "Jewish Racism" and the self-published book *The Jewish Genocide of Armenian Christians* in which he argues that the Turkish killings of Armenians was the result of Jewish design in order to destabilize the Turkish regime, which had control of Palestine.

> Agitated by Jews and crypto-Jews, who hated Christians, the Sultan retaliated against innocent Armenians who were blamed for allegedly stealing the wealth of the Kingdom—wealth which had been stolen by Jewish financiers. These attacks on innocent Armenians benefitted Jewish financiers by weakening an ancient Christian enemy in the region, one associated with the mythical exile of the lost ten northern tribes of Israelites and one associated with the Christians in Jerusalem and elsewhere in Palestine, which Christians then outnumbered Jews in Palestine.[24]

This, he argues, though, was not an isolated evil deed but part of a larger intentional pattern.

> For many centuries, Jews have been fomenting wars of extermination and genocidal revolutions in the nations of the world. They have desperately and successfully sought to bring England, Russia, and the Turkish Empire into world war. Jews have weakened all of the major powers through war and debt in order to dissolve them through revolution and replace them with pan-Jewish empires, as happened with the Roman Empire, the Holy Roman Empire, the Turkish Empire, the Soviet Empire, and the Nazi Empire.
>
> Jewish leaders employ many means to destroy Gentile nations. They foment wars in order to weaken societies and bring them into debt and physical ruin. They conduct strikes and defame leaders and governments

in order to weaken societies and in order to drive a wedge between Peoples and their governments.

Jewish leaders artificially created the Crimean Wars, the Balkan Wars and the World Wars in order to wrest Palestine from the Turkish Empire, destroy the Slavs and Moslems, and discredit Gentile government to the point where Gentiles would welcome a Jewish-ruled world government which promised peace in exchange for the loss of national sovereignty. Jews then mass-murdered Gentiles in the pan-Jewish Empires the Jews created for themselves in the Turkish Empire, the Soviet Empire, and the Communist Chinese Régime.[25]

This line of argument continues in another of his self-published books, *The Manufacture and Sale of Saint Einstein*, where Bjerknes argues

> Adolf Hitler and Joseph Stalin were both agents of the Jewish bankers and both performed the valuable services of segregating the Jews and increasing Jewish hatred of non-Jews. Hitler and Stalin, who were both Bolshevik Zionists, brought the German People and Russian People into war with each other, and helped the Jewish bankers to discredit and ruin Gentile government and to move the world towards a universal world government led by Jews—towards the "New World Order" or "Jewish Utopia" prophesized in *Isaiah* 65:17 and 66:22.[26]

This agenda was served by having an international celebrity who would function as a Zionist spokesman and fundraiser and thereby the myth of the Saint Einstein, the unassailable genius, was born. Despite "all the humiliating defeats Einstein suffered in the scientific world, a pro-Einstein press stuck by him and unfairly smeared those who legitimately criticized him."[27] Einstein was a creation of the public relations wing of the Jewish world conspiracy bent on global domination.

And so one of Bjerknes's life missions is to do the destructive work that needs to be done to save the world, namely, destroying the power of the Einstein mythology, by undermining its source, the belief that Einstein deserves credit for the theory of relativity. In *Anticipations of Einstein in the General Theory of Relativity* and his most famous work, *Albert Einstein, the Incorrigible Plagiarist*, Bjerknes has gone to incredible lengths to document where the history of ideas includes concepts contained in or surrounding the theory of relativity, showing that versions of these notions predate Einstein's work. His books are garnering mainstream scholarly attention, citing him as a source.[28]

The argument in *Albert Einstein, the Incorrigible Plagiarist* is to consider anything that Einstein may have been given credit for and finding someone else in the history of ideas who may have also had that idea. Bjerknes traces elements of the theory of relativity to Lorentz and Poincaré, figures we know influenced Einstein's thoughts, as well as Woldemar Voight (who published the earliest formulation of the Lorentz transformations), and George Francis FitzGerald (who worked to extend James Clerk Maxwell's theory of the aether[29]). He argues, quite sensibly, that Einstein regularly fails to cite those who influenced his thought and his works, something that is definitely a scholarly no-no.

But Bjerknes is not satisfied to point out this misdeed worthy of a severe tsk-tsk. His goal seems to be the obliteration of any sense that Einstein deserves any credit for anything more than trivial application of the theory. With respect to the special theory of relativity, Bjerknes undertakes the task of showing that each element was presented by someone somewhere before Einstein's 1905 paper. His line is that you do not deserve credit for making the salad if you did not grow the lettuce and pick the tomatoes yourself.

In reality, of course, discoveries are never ex nihilo, that is, from nothing. Theories emerge when the time is right, when the ideas have been around, bumping up against each other for a while and someone clever realizes how to join them, tweak them, and augment them to help solve some problem.

Newton's laws were not hatched like Athena emerging from the head of Zeus. There were variants present in Descartes's writings of a generation earlier. But with a radical reconceptualization, these laws were incorporated into a new structure. That's why we think of it as Newtonian mechanics, not because he came up with the theory wholesale, but because he put it together and articulated its picture of the universe for us.

The theory of classical electrodynamics is based around four equations universally referred to as "Maxwell's equations." Of those four equations named for Maxwell, how many did he discover? None. Carl Friedrich Gauss provided two—one describing the electric field around electrically charged particles and one describing magnetic fields around those things that are magnetized. Another describes how a changing magnetic field gives rise to an electrical field and was posited by Maxwell's teacher Michael Faraday. The last comes from André-Marie Ampère who showed the converse, that is, how a changing electrical field creates a magnetic field. Maxwell did make one correction to Ampère's law, but none of the laws named after him was his in the sense that he created them out of nothing.

So, is Maxwell also a fraud to be exposed? Like Newton and Einstein, Maxwell is rightly celebrated because of his synthesis of preexisting bits of knowledge into a unified structure through which we can make sense of the universe in a new way.

The phrase "Copernican Revolution" has come to mean a radical shift in worldview.[30] What Copernicus did was not to invent the sun-centered picture of the solar system but rather to express it in a way that made us see the world around us in a different way. Discovering the new relations among variables is important, but synthesizing them into a new system that changes our concepts, into a form that clearly enunciates a novel and fruitful means of carving up reality, is a valuable act as well, one that does deserve celebration.

But, of course, Einstein at first didn't completely understand the worldview that came from the special theory of relativity. It was Minkowski's geometric interpretation of the theory that took us to the next step. And so, Bjerknes must attack this as well. Minkowski shows us that by considering the invariant and covariant elements of the theory we need to think not in terms of Newton's distinct absolute space and absolute time but an inextricably entwined four-dimensional space-time. What is important is not that space and time are thought of as a single thing, but the particular way that they are related and what it means in terms of the frame dependence of different things we can measure.

For Bjerknes, the jig is up if related ideas can in any way be shown in the works of others. And thus begins an intellectual version of the Kevin Bacon game where tying any thinker from classical Greece onward to any aspect of Einstein's theory is taken as evidence that Einstein deserves no credit for his work. In the case of Minkowski's geometric interpretation, the idea that we live in a dynamic four-dimensional universe with the semi-Riemannian geometry of special relativity should not reflect at all on Einstein because with respect to a fourth dimension, Bjerknes argues, "H.G. Wells, in 1894, expressly stated [the existence of a fourth dimension] in a popular novel, *The Time Machine*, long before Minkowski claimed priority."[31]

Indeed, he contends, the ideas underlying the theory of relativity were already out there in Greece in the fifth century BCE:

> As Dean Turner noted, "space-time," as a concept, as a quadri-dimensional statue, harkens back to the ancients, to Parmenides and the Eleatics, "For what is different from being does not exist, so that it necessarily follows, according to the argument of Parmenides, that all things that are one and this is being." When Minkowski, in 1908, uttered the infamous words, "Henceforth space by itself, and time by itself, are doomed to fade away into mere shadows, and only a union of the two will preserve an independent reality," his words were not only unoriginal, they were trite and more archaic, than arcane.[32]

Parmenides was a pantheist who argued that the only thing that truly exists is what he termed "the One." Classical Greek philosophy makes frequent use of the notion of "teleology," that all change is goal directed. What changes, therefore is imperfect, since it is changing toward something. What is truly real, Parmenides argued, is that which has full realization, that which is unchanging. But since change is goal directed it means anything that can change is imperfect and therefore unreal, merely apparent. All that there is is the underlying reality, the One. So, Parmenides united the entire universe into a single unchanging entity, and Minkowski showed that Einstein's theory unites space and time so that change can be understood according to equations that tell you how measurements of individuated objects will be affected when moving from one reference frame to another. You say tomato . . .

But Bjerknes's goal is clear. There must be nothing left of Einstein's reputation and value as a cultural figure. Einstein's work, Bjerknes wants to argue, is Jewish science.

A THIRD EXAMPLE OF A CONTEMPORARY WRITER who objects to Einstein's work as "Jewish science" is William Lane Craig. Unlike others in this category, he argues that the evidence supports the theory and that the conversation should move away from personalities or politics. He is the René Descartes of contemporary evangelical Christian philosophers of physics. Just as Descartes labored to reconcile the papal decree that the Earth does not move with the Copernican sun-centered picture of the universe, so too Craig works hard to make the theory of relativity mesh with a standard understanding of contemporary Protestant theology.

The Protestant God as a "transcendent, personal Creator of the Universe" and the theological ramifications of such a Being, he argues, hinge on notions that make inextricable use of the idea of time. If God is the source of all that is, then there must be a meaningful sense of coming-to-be. But coming-to-be is a temporally loaded notion, implicitly assuming a now and an "is later than" relation between all events in time, that is, there must be a sense in which all happenings are completely ordered. This, Craig argues, requires the concept of absolute time that Newton championed in his "Protestant" physics.

We humans may be limited to a relative reference frame in which we can consider measured physical time, but for God to exist in the universe is for God to exist in time. With the notions of divine foreknowledge and Creation, it must follow that there is one reference frame that is privileged, that is, God's metaphysical time must exist.

> Poincaré . . . suggests that, in considering the notion of simultaneity, we instinctively put ourselves in the place of God and classify events as past, present or future according to His time. Poincaré does not deny that such a perspective would disclose to us the true relations of simultaneity . . . The fact remains that God knows the absolute simultaneity of events even if we grope in darkness.[33]

There must be a real "now" in which God resides for there to be creation. God is eternal, but he is not, Craig argues, timeless. A present God is a temporal, albeit eternal, God.

But the notion of time underwent a radical revision with the theory of relativity. Making simultaneity and time order frame-dependent as Einstein did runs counter to the theology Craig assumes as a starting point. Yet, he cannot deny that this frame-dependence is observed, quoting Hermann Bondi: "There is perhaps no other part of physics that has been checked and tested and cross-checked quite as much as the Theory of Relativity."[34] Craig needs to find a way to have his relativistic cake and still eat it in Newtonian absolute time, too.

Where Descartes had to create a whole new theory to accomplish his version of the task, Craig inserts God into Minkowski's united space-time by creating a new theory from the old one by reinterpreting Einstein's theory.

> A physical theory is comprised of two components: a mathematical formalism (a set of equations and a set of calculational rules for making predictions that can be compared with experiment) and a physical interpretation (what the theory tells us about the underlying structure of the phenomena, that is to say, an ontology). Thus, a single formalism with two different interpretations counts as two theories.[35]

Einstein gave us equations and Minkowski gave us his geometric interpretation that provides a story about what a world governed by those equations looks like, what philosophers call an "ontology." What Craig seeks to do is to find a new story, a different narrative about the universe that keeps those equations functional, but that retains an absolute time.

He does this by adding a new time underneath of time. Craig distinguishes between "physical time" measured by clocks and "metaphysical time," which is the real time in which God resides. This distinction is nothing new. Think back to Newton, who, in seeking to undermine Descartes's notion of a movable space and relative motion, sets out his absolute concepts of space and time distinguishing his notions that were true, absolute, and mathematical from Descartes's that were merely apparent, common, and relative. Indeed, that these concepts also had theological foundations in Newton's Protestantism is a point not lost on Craig.[36] This time underneath of time is, then, the absolute time of Newton. But the laws that govern the motions of objects in that time appear from the human perspective to be those of Einstein's theory. Minkowski's work is a mere tool.

> Minkowski's four-dimensional mathematical space serves as a convenient calculational and diagrammatical aid, a mathematical *Hilfsmittel* [resource], but says absolutely nothing about ontology.[37]

This is the same sentiment we find in Andrew Osiander's introduction to Copernicus's posthumous masterpiece *On the Revolutions of Heavenly Spheres* that argues that Copernicus's heliocentric view does not warrant the hubbub it created as it too is solely a calculational aid, not a picture of how the world really is.

> For it is the job of the astronomer to use painstaking and skilled observation in gathering together the history of the celestial movements, and then since he cannot by any line of reasoning reach the true causes of these movements—to think up or construct whatever causes or hypotheses he pleases such that, by the assumption of these causes, those same

movements can be calculated from the principles of geometry for the past and for the future too.[38]

In both cases, the physical equations allow us to do powerful work making predictions, but do not give the true underlying nature of reality. For that, Craig argues, we must turn to theological considerations. Equations make predictions but cannot give you a glimpse of what lies beneath.

This move of creating two different times, one of human experience and another of God's perspective, has tangible ramifications for Craig in making traditional theology and modern physics coherent. It allows us to talk about the Big Bang as the moment God created the universe.

> P.C.W. Davies, observing that if a spacetime singularity did occur at the Big Bang, as predicted by the Friedmann models, then it will be impossible "to continue physics, or physical reasoning, through it to an earlier stage of the universe," goes on to conclude from this that it is 'meaningless' to speak of god's creating the universe. For a cause must precede its effect temporally, but there is no temporal moment before the Big Bang. Therefore, the Big Bang can have no cause. But (leaving aside the premise that a cause must temporally precede its effect) if we draw a distinction between metaphysical time and physical time as Newton did, it is quite evident that a beginning of the latter does not imply a beginning of the former. God in metaphysical time could be quite active prior to creation (perhaps creating angelic realms) and could bring physical space and time into being after having existed without their being co-existent with Him.[39]

By creating a second time, we can embed the Big Bang with its inferior limited physical time in the eternal and external metaphysical time of God.

In this way, we see another objection to Einstein—if not to his physics, at least to his understanding of the universe that he drew from his physics—based on religious grounds. Craig gives us a Christian physics, albeit one with Einstein's equations that is explicitly distinguished from Einstein's own science. In a number of different ways, the contention that there is something wrong with the theory of relativity because it is in some sense "Jewish science" lives on.

CONCLUSION

Einstein's Cosmopolitan Science

So, in the end, is Einstein's theory of relativity Jewish science or isn't it? In most of the ways we've examined, it isn't. But the Nazis sure were bothered by it. And well they should have been. Einstein's science *is* in deep and interesting ways a challenge to the heart of their ideology and in as far as the Nazis held "Jewish" to be the antithesis of their own view, the approach to universe that we inherit from Einstein's development of the theory of relativity does undermine their way of making sense of the world.

Einstein came to the scientific stage at a time when Western culture was in flux. Old social, political, economic, artistic, and intellectual structures were falling. Assumptions that had been protected for centuries were suddenly rejected despite all attempts to maintain them. And here, offering a new and bizarre way to see the entire universe was Einstein. The theory of relativity stands as a symbol of Gestalt shift, a complete change in perspective where you can never see the familiar in the same old way.

Einstein was vilified by those who clung to the old order. His science, his politics, and his views about religion were all made public in ways that made them difficult to ignore. Einstein once joked that he had a strange version of the Midas touch, everything he said turned to newsprint. And more often than not, what was printed was not what those who wanted things to stay the same wanted to hear.

He was elevated, in some cases willingly and in others not—to the position of spokesperson against the old ways, and for this he was hated.

But for those who saw the change as a part of human progress, of cultural growth, he was part of the engine driving us forward. His science was revolutionary, but it wasn't just his science. Einstein, the man, as much as the theory of relativity, became the symbol of this new way. He eschewed belts and socks. His hair, that iconic hair, stoked a sense of nonconformism to that which is merely social construction. Here was a great mind who rejected the trivial. Think of the quotations that one most often sees attributed to Einstein on bumper stickers: "Imagination is more important than knowledge." "The only thing more dangerous than ignorance is arrogance." "Great thinkers have always encountered violent opposition from mediocre minds." What we celebrate about Einstein is his embrace of a certain form of Enlightenment values: that humanity progresses when it thinks creatively, when it stops accepting culturally enforced strictures and frees itself in the search for truth. We take Einstein to be the epitome of the open mind.

Of course, this sort of icon-making is a form of mythologizing. Einstein was not the intellectual or moral saint he is portrayed to be. But his theories, actions, and words appeared at a time when things looking forward were going to be radically different from things looking backward. And at the heart of it all was relativity.

Physics occupies a special place in the cultural imagination. We live in a universe where we can think of ourselves as the center of Creation or we may think of the universe as bigger than us, stranger than we can imagine, and full of mystery. Some desire the mysteries to remain forever mysterious and for them, the threat of science exposing underlying order is worrisome. In this way, Einstein is a threat.

But for those who relish the mysteries that emerge *from* the understanding, who see each step forward as opening new, more wonderful and magical ways of being in the universe, Einstein's picture of reality seems to resemble the works of M. C. Escher in which the oddity of the image is seen as a virtue. If the universe is strange, then we must be as well. Our minds must bend in new and exciting ways if we are to be at home in this new world.

This openness toward the way we think of the universe, the government, and everything on down to the way we dress, is a form of liberation. And Jews see the emblem of this new era of openness as one of their own. The liberation Einstein represents is in part Jewish liberation. Einstein still resonates as a symbol of intellectual freedom in a generation when Jews no longer have significant worries about anti-Semitism hampering social mobility or personal opportunity. The idea that one needs to convert to Christianity for social advancement seems absurd today. Clubs and housing are no longer restricted. University appointments are no longer a question. In so many ways, Jews are completely assimilated into American culture while still being able to maintain a personal identity that remains apart from the rest of the society.

Einstein stands as a member of an ancient people, yet creating a new future for all of us. Einstein represents the way in which Jews, whether they are secular or religious, remain relevant in a world in which science and technology are reshaping every aspect of life. Judaism is inherently fluid, constantly redefining itself, in some ways allowing old elements to drop off in assimilation, but—and this is where Einstein acquires some of his meaning—also changing the greater cultural landscape.

With Jews no longer in the ghetto—physically, socially, or intellectually—Einstein represents a third way that stands opposed to both segregation and assimilation. He represents cosmopolitanism. What makes the theory of relativity so important to Jews is precisely that it is *not* Jewish science.

The old structures that crumbled after World War I were in deep ways the ultimate expression of the power of Christian culture since the Renaissance. The new social, political, and intellectual structures being built now to replace them in the "new world order" that so threatens contemporary conservatives are not Jewish but are constructs within which Jews will find something familiar and not oppressive. The theory of relativity is in most senses *not* Jewish science, but it is a symbol of this new way of being in the world. Indeed, it literally creates a new universe for us all to occupy, one that is consistent with ways of approaching the world that one finds in Judaism historically. That familiarity and the ability to be a vital part of the future it portends is why Jews love Albert Einstein. He helped open the door to a new world and in that strange realm stretching out before us, one can be a full participant *and* be Jewish.

But this love of Einstein is broader than that of the Jewish community, because the liberated, cosmopolitan future he represents is not at all tied to a connection between Judaism and relativity. Einstein was a secular Jew, but he very well could have been a religious Jew like Nobel laureate I. B. Singer; he could have just as well been a Hindu like Nobel laureate Chandrasekhara Venkata Raman; he could have been an Arab like Nobel laureate Ahmed Zewail, he could have been a woman like Nobel laureate Marie Curie; he might have been gay like mathematical genius Alan Turing. Einstein is Einstein because he says to members of every oppressed group that he could have been like you. His theory, his irreverence, his politics, and the opposition he faced represent an opening up and destruction of the confining strictures that limited whose voice could be heard, in what language it must be spoken, and with what accent. His place in history was a pivotal moment at which the future no longer resembled the past. That future will no longer be dominated by the powers that entrenched themselves over the past several centuries and Albert Einstein is a symbol that all of us can participate in forming the future.

We all love Einstein because he is Jewish, not because Judaism is to be privileged in any way, but because it was marginalized, indeed in Einstein's times, it is the very epitome of marginalization by the powerful. Einstein showed us a new world in which space and time are not rigid and absolute but frame-dependent. A strange new realm that challenges the "commonsense" notions we were indoctrinated with. He shows us that we can live in that bizarre place and can do so comfortably without belts or socks, wearing our hair in any way we please, while making jokes and sticking out our tongues. The Nazis who argued that Einstein's Judaism is part and parcel of a false refinement that was undermining a truly authentic posture towards the world had it completely backward. It stands as a challenge to refinement and the denial of so many authentic ways of being, all of which contribute to building the world and our understanding of it. It is not a homogenization, but a come-as-you-are party in which absolute truths *do* exist, but which all of us can, with diligence, hard work, and international influences, contribute to finding and understanding in deeper and broader ways.

But while Einstein's scientific results are the quintessence of the new modern worldview, and while Einstein the icon, if not Einstein the man, is a universally recognizable symbol of the new liberation that accompanies the new ways, there is still something deeper in the reasoning process surrounding the theory that is cosmopolitan in a deep way that threatens the sort of narrow-minded provincialism hyper-exemplified by Nazism.

The method labeled as "Jewish-style" reasoning in which an absolute truth exists, but is unavailable to any particular frame of reference, provides the groundwork for what we might call a "cosmopolitan epistemology," that is, a new way of treating human knowledge in which we take seriously different perspectives, but do so in a fashion that does not undermine our belief in a real world or the human ability to develop objective truths about it.

We called this approach metaphorically Jewish or Jewish style because there is something formally similar in Talmudic interpretation, in which a larger metaphysical truth, God's truth, is presumed to exist, but which is as a whole beyond our capacity to grasp, but elements of which are illuminated, uncovered, exposed, by seemingly contrasting human interpretations. When we move from these metaphysical truths to the workings of the material universe, we see a version of this style of approach still applicable, but now the perspectives are not interpretations of scripture. In Einstein's physics, they become measurements from different frames of reference. But we can consider Einstein's approach extended to other intellectual contexts as well, creating a general cosmopolitan approach to knowledge.

What is fascinating and heartening is that we see this exact notion being developed in parallel fashion by thinkers from other marginalized groups. Emmanuel Chukwudi Eze, among the most prominent contemporary postcolonial Africana philosophers, in his last book before a tragic, premature passing, wrote on the notion of reason in a world of different groups with different histories and resources.

> To adequately comprehend what we are thinking when we say that someone is behaving rationally, and arrive at such comprehension from the legacies of multiple traditions in philosophy and from insights available in the nearby disciplines, requires complex levels of exploration. But, above all, it is my hope that, taken together, the various practical intuitions, interpretations, and explorations, and justifications will constitute different examples of reason-in-act reveal the grounds of reason's abstract principles as well as the relations of these principles to practical interests of individuals, institutions, and cultures. I thereby hope to discover how to test not just the claims of individuals to rationality but also the rationality of cultural practices and other kinds of events that claim to be productive of reason. My methodology, indeed, provides proof of the truth of the statement *Anaghi akwu ofu ébé ènéné manwu*: Rationality, like a work of art, is best appreciated from multiple points of view.[1]

Eze is working out a cosmopolitan epistemology along the sort of path as we see in Einstein.

For Eze, thought requires diversity. By that, he means that thought begins with language—"Thought is the need for language"[2]—but language is laden with culture and history. All languages contain the views and biases of the linguistic community that speaks it and those categories are then implicit in the ways we think.

Languages may contain key elements of worldviews, but is not enough by itself to completely determine our thoughts. There are gaps, what Eze terms "breaches in the tongue," and these spaces are the places where individual thought and freedom emerge. In being a part of a linguistic community, we are constrained, but also find pockets of liberation. We are a part, but also apart. Thinking requires this diversity within the mind.

But the notion of diversity is broader still for Eze. Since thought, and therefore rationality, stem from the language, and since humans have developed multiple languages, there will necessarily also be multiple rationalities.

> It is because language is thus thoroughly historical that thinking, too, is historically fated. And inasmuch as thinking is both worldly and historically fated, there cannot be just one way or one kind of expression of thought. There are many forms of expressions of thought. There are many universal languages of reason.[3]

The Enlightenment concept of a single sort of rationality runs into the problems we saw with David Hume. The additional claim that there is a single perspective which achieves this rationality, what we have seen as a "Christian-style" approach with a God's eye perspective that humans can occupy, has its own additional problems. But if we allow there to be "many universal languages of reason," each tied to a historical and cultural reference frame, then we get a richer, more Africana—and seemingly more "Jewish-style"—approach to reason. Just as the Nigerian-born, but Fordham-educated Eze did, we must use different "languages of reason" to illuminate the universal Truth from different perspectives, each allowing us to glimpse it more deeply and thoroughly than we can from a single limited viewpoint.

We see the same sort of move in the writings of Sandra Harding, a prominent feminist philosopher of science, who champions what she calls "strong objectivity" when tied to standpoint theory. The traditional picture of reason, the "Christian-style" approach, has a privileged viewpoint from which all truth is accessible. But, of course, all human perspectives are limited and warped. We remain blind to this distortion because it is normal; it is the way we have always made sense of the world. This is as true in scientific research as it is in everyday life.

> Our cultures have agendas and make assumptions that we as individuals cannot easily detect. Theoretically unmediated experience, that

> aspect of a group's or an individual's experience in which cultural influences cannot be detected, functions as part of the evidence for scientific claims. Cultural agendas and assumptions are part of the background assumptions and auxiliary hypotheses that philosophers have identified. If the goal is to make available for critical scrutiny *all* the evidence marshaled for or against a scientific hypothesis, then this evidence too requires critical examination *within* scientific research processes.[4]

Cultural elements infect our thoughts in ways we don't realize and this not only shapes, but constrains what we can know.

> The assertion is that human activity, or "material life," not only structures but sets limits on human understanding: what we do shapes and constrains what we can know. As [Nancy] Harsock argues, if human activity is structured in fundamentally opposing ways for two different groups (such as men and women), "one can expect that the vision of each will represent an inversion of the other, and in systems of domination the vision available to the rulers will be both partial and perverse."[5]

As a result, there is an advantage to being in the out-group because while you are trained by members of the in-group, you also have access to experiences and perspectives that the others do not.

> The feminist standpoint theories focus on gender differences, on differences between women's and men's situations which give a scientific advantage to those who can make use of the differences.[6]

These advantages are contextual. Does it affect astronomers? Not so much. Biologists? Sure. Sociologists and anthropologists? Absolutely.

This might lead some to adopt a completely perspectivalist line, like the postmodernist thinkers, according to which we have to deny the existence of objectivity altogether. But Harding moves in a different direction, arguing that objectivity is essential for science. There is a world and we can have objective knowledge of it. But since we know that our perspective is limited, a "strong" sense of objectivity is one in which—like Einstein—we need to find what is true about the larger truth that is too big to fit in any one perspective. We use differences to adjust for biases, to correct for perspective-induced errors. There are truths from standpoints, but then there are the objective truths across standpoints and the goal of science is to find these larger truths.

So, from feminist and postcolonial Africana thinkers, we find the same sort of approach that we found in Einstein and classical Jewish thinkers—that there is an absolute truth, but it is not accessible from any special viewpoint, rather we can glimpse it if we understand how views change across different perspectives. It is interesting that in all of these cases, a similar approach to knowledge emerges from those who have been oppressed, but yet who are currently in the process of making their mark on intellectual history.

What is important here is not dwelling on the oppression—although certainly we need, in all cases, a full historical accounting. What gives Einstein his power is the second element, that this is a view about hope, liberation, and finding truth. Einstein, in giving us an example of this approach to knowledge, exemplifies the promise that it could belong to any of us. The cosmopolitan epistemology brings with it a sense of human progress. We can make life better. We can have a better understanding of the universe around us. But this progress is undermined if any of us consider ourselves the master race, or even God's Chosen People.

So, is Einstein's theory of relativity Jewish science? Yes and no. And *that* is precisely what makes it Jewish.

ACKNOWLEDGMENTS

I would like to thank a number of people who helped in the preparation of this book. Larry Marschall, Lisa Portmess, Kerry Walters, Hanno Bulhof, Jennifer Hansen, Gary Mullen, Don Goldsmith, Donna, Rick, and Adam Rosenberg all offered very helpful and insightful comments on earlier versions. This project could never have come into being without the tireless devotion of Deirdre Mullane of Mullane Literary Associates and has been immeasurably improved by Michele Callaghan, my copy editor at the Johns Hopkins University Press,

I would especially like to thank Lou Hammann, not only for his enthusiastic support of this project from its inception but also for a lifetime of work and play, creating an intellectual environment from which this sort of conversation could emerge.

This project began as a scholarly article cowritten with Professor Stephen Stern, who has graciously allowed me to include elements of that original work here. This book would not have been possible without his help and many, many, many hours of discussion concerning the topic. His ideas, objections, and corrections were essential for the development of the arguments and insights contained here. However, any factual errors or controversial claims made in this work are to be attributed solely and completely to the author.

NOTES

Introduction. Einstein's Jewish Science

1. Reichenbach, "Die Einsteinische Bewegungslehre," *Die Umschau*, vol. 27 (1921), p. 503. Translated as "Einstein's Theory of Motion," p. 73. For a range of reactions to the theory, positive and negative, see Raman (1972).

2. Lenard, *Deutsche Physik in vier Bänden: Einsteilung und Mechanik*. Foreword translated as *German Physics* in Hentschel, *Physics and National Socialism*, p. 100.

3. Einstein, "Anti-Semitism: Defense through Knowledge," *Collected Papers*, vol. 7, p. 294.

4. Merton, *Science, Technology, and Society in 17th Century England*.

5. Weizmann figured out how to use bacteria to break down compounds in a way that could produce large amounts of desired derivative chemicals. Industrial chemical production before Weizmann focused on chemical reactions, but Weizmann realized that biological agents could be the key to safely and quickly producing mass quantities of useful chemicals. This discovery led to the enhancement of British arsenals necessary for fighting the world wars.

6. Einstein, "How I Became a Zionist." *Collected Papers*, vol. 7, p. 234.

7. Quoted in Rowe, "Jewish Mathematics," p. 424.

8. Bosons are particles whose behavior is governed by mathematical relations called Bose-Einstein statistics that were developed initially by Satyendra Bose and expanded on by Einstein. Particles are bosons if they can exist side by side with other particles in an identical state, what physicists call "gregarious" particles. Electrons, like Einstein, turn out to be loners and so are not bosons.

9. See, for example, Fuller, *Science*, or Latour and Woolgar, *Laboratory Life*.

Chapter One. Is Einstein a Jew?

1. See Satlon, "Wasted Seed," pp. 158–62, and Boyarin, *Unheroic Conduct*, p. 9.

2. For a detailed discussion of the different proposed explanations of the origin of the matrilineal principle, see Cohen, *Beginnings of Jewishness*, pp. 283–303.

3. July 29, 1900. Reprinted in Renn and Schulmann, *Albert Einstein / Mileva Marić*, p. 19.

4. Reprinted in Dawidowicz, *Holocaust Reader*, p. 39.

5. "First Decree for Implementation of the Law for the Restoration of the Professional Civil Service, April 11, 1933," reprinted in Dawidowicz, *Holocaust Reader*, p. 41.

6. Ibid.

7. Ibid., p. 46.

8. Ibid.

9. Many German eugenicists were opposed to Nazism and saw eugenics not as a racist means of promoting one group over another but as a means of improving the general health, and therefore welfare, of all of humanity. Some saw the rise of Nazism as a boon to their research, since the Nazi fascination with eugenics would mean public funding, but others were quite concerned about the connection. For discussions of Nazism and eugenics, see Kühl, *The Nazi Connection*, Weikart, *From Darwin to Hitler*, and Hentschel, "Bernhard Bavink (1879–1947)."

10. For discussions of medical research under the Third Reich, see Proctor, *Racial Hygiene*, and Weindling, *Health, Race and German Politics*.

11. Dawidowicz, *Holocaust Reader*, pp. 46–47.

12. See Mommsen, *From Weimar to Auschwitz*, p. 229, and Hilberg, *Destruction of the European Jews*, p. 47.

13. Hilberg, *Destruction of the European Jews*, p. 49.

14. Davidson, *Making of Adolf Hitler*, pp. 5–6.

15. Hilberg, *Destruction of the European Jews*, p. 48.

16. Kuhn, *Structure of Scientific Revolutions*.

17. Clark, *Einstein: The Life and Times*.

18. Two nice biographies of Meitner are Rife, *Lise Meitner and the Dawn of the Nuclear Age*, and Sime, *Lise Meitner: A Life in Physics*.

19. See Feuer, *Einstein and the Generation of Science* for a discussion of the ways in which Swiss life influenced Einstein's thought, especially pp. 4–13.

20. "Letter to the Central Association of German Citizens of the Jewish Faith" April 5, 1920, in *Collected Papers of Albert Einstein*, vol. 9, p. 368.

21. Einstein, "Intellectual Autobiography," in Schilpp, *Albert Einstein: Philosopher-Scientist*.

22. Ibid.

23. Ibid., p. 9.

24. Letter to M. Berkowitz, Oct. 25, 1950. Quoted in Calaprice, *The Expanded Quotable Einstein*, p. 216.

25. See Hume, *Treatise of Human Nature*, book III, "Of Morals."

26. From a letter to a rabbi in Chicago; Einstein, *Albert Einstein: The Human Side*, pp. 69–70.

27. Einstein, "Religion and Science," *New York Times Magazine*, Nov. 9, 1930, reprinted in *Ideas and Opinions*, p. 39.

28. For an extended discussion of Einstein's "Cosmic Religion," see Jammer, *Einstein and Religion*.

29. Rebecca Goldstein, in her book, *Betraying Spinoza: The Renegade Jew Who Gave Us Modernity*, argues that the Spinoza's views are not in fact heretical, as was—and still is—claimed, but rather are much more Jewish than Spinoza is given credit for.

30. Spinoza, *Ethics*, p. 10.

31. "Einstein believes in 'Spinoza's God,'" *The New York Times*, April 25, 1929, p. 60.

32. Einstein, *Ideas and Opinions*, p. 11.

33. Quoted in Feuer, *Einstein and the Generation of Science*, p. 80. Feuer notes that in the otherwise excellent translation of the Bohr-Einstein correspondence by Irene Bohr, the phrase "der wahre Jakob" is rendered "the real thing" depriving it of its biblical allusion.

34. Pais, *Subtle Is the Lord*, p. 9.

35. Einstein, "Let's Not Forget," reprinted in *Out of My Later Years*, p. 256.

36. Einstein, "Defining Judaism," p. 853.

37. Einstein, *Ideas and Opinions*, pp. 185–86.

38. For an in-depth discussion of Einstein's views on American racism and his relationships with Du Bois and Paul Robeson, see Jerome and Taylor, *Einstein on Race and Racism*.

39. Einstein, *Out of My Later Years*, pp. 132–33.

40. See Jerome, *The Einstein File*.

41. Einstein, *Autobiographical Notes*, p. 5.

42. Einstein, *Ideas and Opinions*, p. 186.

43. Einstein, "Intellectual Autobiography," p. 9.

Chapter Two. Is Relativity Pregnant with Jewish Concepts?

1. For an extended discussion of catastrophism and its place in the history of geological thought, see Gohau, *A History of Geology*.

2. For a comparison of classical astronomical thought from around the globe, see John North's extremely thorough work *Cosmos*.

3. Comte, *Positive Philosophy*.

4. We often use the phrase Judeo-Christian instead of Judeo-Christian-Islamic, which wrongly ignores the third Abrahamic tradition. While this use is meant to be more inclusive, it is also historically essential to include Islamic thought. Muslim astronomers in the Middle East and Spain were responsible for major steps forward in cataloguing and understanding the motions of the heavens. These steps were further advanced by Christian astronomers. In both cases, one of the driving motivations was a combination of religion and time—Muslims needing to determine the start of Ramadan and Christians needing to determine the date of Easter. In classical times, astronomical observation was the only way to determine these temporal features exactly. This led to a strong interest in astronomy on the part of the religious communities. See North, *Cosmos*, for detailed discussions.

5. A number of biographies of Descartes have been written recently. See, for example, Grayling, *Descartes: The Life and Times of a Genius*; Gombay, *Descartes*, and Clarke, *Descartes: A Biography*.

6. For a wonderful account of Galileo's life and relation to the Church, see Sobel, *Galileo's Daughter.*

7. Galileo, *Dialogue Concerning Two Chief World Systems.*

8. Descartes, *Principles of Philosophy.*

9. For accounts of the life of Isaac Newton, see Manuel, *A Portrait of Isaac Newton*; Westphal, *Never at Rest*; White, *Isaac Newton: The Last Sorcerer*; Christianson, *Isaac Newton*; and Gleick, *Isaac Newton.*

10. Newton thought Hermes Trismegistus to be the greatest of the ancient practitioners of alchemy, not an unusual belief for the time. Others, though, considered him an Egyptian god, while still others considered it a fictional name used by alchemists through the centuries for their own treatises in order to give their work credibility by trading on the famous appellation.

11. Newton, *Mathematical Principles of Natural Philosophy.*

12. Newton, *Principia,* p. 544.

13. Letter to Richard Bentley, Dec. 10, 1692. *Isaac Newton's Papers,* p. 280.

14. Newton, *Opticks,* p. 370.

15. Newton, *Principia,* p. 545.

16. Ibid., p. 547.

17. Letter to Richard Bentley, Jan. 17, 1693. *Isaac Newton's Papers,* p. 298.

18. Ibid.

19. For a full discussion of Einstein's influences in the development of the special theory of relativity, see Miller, *Albert Einstein's Special Theory.*

20. Einstein, "The Relative Motion of the Earth in the Ether" in *Collected Papers,* vol. 4, pp. 219–23. Quoted in Miller, *Albert Einstein's Special Theory,* p. 31.

21. Mach, *Science of Mechanics,* p. 279.

22. For a discussion of Poincaré's views and their influence on Einstein's work, see Holton, *Einstein's Clocks, Poincaré's Maps.*

23. Kant's masterwork *Critique of Pure Reason* sets out an intricate account of all of human knowledge—scientific, mathematical, observational, metaphysical, ethical—in a way that brought together aspects of the philosophy that came before and shaped virtually all Western philosophy that followed.

24. Poincaré's conventionalist view is most famously worked out in his book *Science and Hypothesis.*

25. Poincaré, "The Measure of Time" in *The Value of Science,* pp. 27–28.

26. Reprinted in Lorentz et al., *Principle of Relativity.*

Chapter Three. Why Did a Jew Formulate the Theory of Relativity?

1. Of course, that did not stop any number of writers from making such speculations. Some do attribute Einstein's genius to features inherent in Jewish culture, most notably Thorstein Veblen in his 1919 "The Intellectual Pre-eminence of Jews in Modern Europe." Others, such as Norbert Weiner, argue that such genius is a biological feature that was evolutionarily selected for because Jews, unlike Catholics, allow rabbis to have children. Cause-and-effect arguments of any sort, whether

nature-based or nurture-based, on such questions necessarily overshoot their bounds. Even if one could cite autobiographical evidence, one always has to worry about the legitimacy of such reconstructed narratives.

2. Einstein, "Anti-Semitism: Defense through Knowledge," in *Collected Papers*, vol. 7, p. 294.

3. Descartes, *Discourse on Method*, p. 4.

4. Euclid, *The Thirteen Books of Euclid's Elements.*

5. Titus, 2:5-7 (New Revised Standard Version).

6. Loeb, *Apostolic Fathers*, vol. 1, pp. 201–3.

7. For a fascinating discussion of the context, working, and ramifications of the Council of Nicea, see Rubenstein, *When Jesus Became God.*

8. *Confessions of St. Augustine*, p. 70.

9. Plato, *Republic.*

10. Bacon, *Novum Organon.*

11. Newton, *Principia*, pp. 398–400.

12. William of Ockham was a fourteenth-century English thinker who wrote on the nature of existence. In such reasoning, he contended, one should not multiply entities beyond necessity, a later statement of his famous principle of parsimony.

13. Newton, *Principia*, p. 399.

14. For a history of the Medici, see Hibbert, *The House of Medici.*

15. On the life of Luther, see Marty, *Martin Luther: A Life.*

16. *Martin Luther: Selections from his Writings*, pp. 496–97.

17. Ibid.

18. "Freedom of a Christian" in *Martin Luther: Selections from his Writings*, p. 55.

19. Sylvester's life is chronicled by Karen Parshall in *James Joseph Sylvester: Life and Work in Letters* and *James Joseph Sylvester: A Jewish Mathematician in a Victorian World.* For Cayley's biography, see Crilly, *Arthur Cayley.*

20. Sylvester, "Note on a Proposed Addition," pp. 152–53.

21. The life of Klein is set out in Yaglom, *Felix Klein and Sophus Lie.*

22. For a discussion of the race between Einstein and Hilbert to develop the general theory of relativity, see Mehra, *Einstein, Hilbert, and the Theory of Gravitation.*

23. The interrelated lives of Hilbert and Minkowski are recounted in Reid, *Hilbert.*

24. Quoted in Clark, *Einstein: The Life and Times*, p. 123.

25. Einstein, "Intellectual Autobiography," p. 15.

26. The quotation appears in Born, *Physik im Wandel meiner Zeit*, p. 218, quoted in Miller, p. 221.

27. Minkowski, "Space and Time" in *Theory of Relativity*, pp. 73–91.

28. Ibid., p. 75.

29. Meyers, "Jewish Culture in Greco-Roman Palestine," p. 149.

30. Note that the Hebrew Bible as we know it today had not yet been codified, so one cannot speak simply of a single canonical text in this context.

31. It is not clear whether the term "rabbi" was used at this time in anything resembling its current meaning.

32. Meyers, "Jewish Culture in Greco-Roman Palestine," 163.

33. For a comprehensive introduction to the history of the Talmud, see Steinsaltz, *Essential Talmud.*

34. *The Essential Talmud,* p. 3.

35. Ibid.

36. Einstein, in *The Theory of Relativity,* pp. 35–65.

37. Einstein, *Einstein's Miraculous Year,* pp. 177–79.

38. While Newton raises this question about light traveling through empty space, he frames it as a question. Later advocates of the Newtonian position, however, saw it as a knock-down argument for the corpuscular theory of light. For an extended discussion of the argument over the nature of light in the seventeenth and nineteenth centuries, see Achinstein, *Particles and Waves.*

39. For accounts of the history of Planck's work on the problem of blackbody radiation that are accessible to the general reader, see Cline, *Men Who Made a New Physics,* and Jones, *The Quantum Ten.*

40. "The Foundation of the General Theory of Relativity" in *The Theory of Relativity,* pp. 109–64. For a wonderfully accessible account of the general theory of relativity, see Geroch, *General Relativity: From A to B.*

41. Freud, *An Autobiographical Study,* p. 12.

42. Ibid, p. 14.

43. "Freud's First Interview on Psycho-Analysis," *New York American.* Aug. 19, 1923.

44. Freud, "Resistances to Psychoanalysis," *La Revue Juive.* Quoted in Lowenberg, "Sigmund Freud's Psycho-Social Identity," p. 144.

45. Freud, *An Autobiographical Study,* p. 13.

46. Freud, *Complete Introductory Lectures,* p. 207.

47. Freud, "Notes upon a Case of Obsessional Neurosis," in *Three Case Histories.*

48. Freud, *An Autobiographical Study,* p. 49.

49. The argument here is not whether Freud was influenced by his Jewish roots. David Balkan in *Sigmund Freud and the Jewish Mystical Tradition* argues that there is a connection between psychoanalysis and the Kabala. This is not what I am arguing for or against here but rather that if you look at the schematic reasoning structure that is present in Talmudic conversation, it is not present in Freud.

50. Durkheim, *Suicide: A Study in Sociology.*

51. Durkheim, *The Rules of Sociological Method.*

52. Ibid., p. 1.

Chapter Four. Is the Theory of Relativity Political Science or Scientific Politics?

1. For a detailed discussion of this period, see Kitchen, *A History of Modern Germany.*

2. See, for example, Weiner, *Richard Wagner and the Anti-Semitic Imagination.*

3. Wagner, "Judaism in Music," p. 83.

4. Ibid.

5. Nietzsche, *On the Genealogy of Morals*, p. 37.

6. Ibid., pp. 40–41.

7. Ibid., p. 32.

8. Ibid., pp. 33–34. Italics in the original.

9. One should not take from these quotations that Nietzsche was a garden variety anti-Semite. Nietzsche also had plenty of venom for Christians, among most others. Indeed, there were very few people Nietzsche liked. One of them was Richard Wagner, whom he held in high esteem but with whom he had a falling out because of Wagner's overt anti-Semitism.

10. Reulecke, "The Battle for the Young," p. 98.

11. For a discussion of these movements, see Wohl, *The Generation of 1914*, and Mitteraur, *A History of Youth*.

12. Mitteraur, *A History of Youth*, p. 213.

13. Ibid.

14. For a discussion of the Nazi nature protection laws, see Ferry, *The New Ecological Order.*

15. Bratton, "Luc Ferry's Critique of Deep Ecology," pp. 3–22.

16. Hegel, *Phenomenology of Spirit.*

17. Marx, *Capital.*

18. For an in-depth discussion, see Mommsen, "The Decline of the *Bürgertum*."

19. Ibid., p. 592.

20. Schopenhauer, *On Human Nature*, p. 57.

21. For a stark portrait of Wilhelm II, see Carter, *George, Nicholas, and Wilhelm.*

22. For a detailed discussion of the birth of the Weimar Republic, see Mommsen, *The Rise and Fall of Weimar Democracy.*

23. Quoted in Kitchen, *A History of Modern Germany*, p. 177.

24. The new liberal government also enraged the far left, who tried to establish Germany as a Lenin-style soviet republic, but the center and center-left democrats put down the Sparticus revolt and the population made it clear in parliamentary elections that they wanted a democratic structure, thereby repudiating both the left and the right.

25. Pulzer, *Jews and the German State*, pp. 217–25, gives a nice discussion of the German Democratic Party.

26. Ibid., pp. 218–21.

27. Fritz Stern's chapter, "Walther Rathenau and the Vision of Modernity" in *Einstein's Jewish World* paints a detailed picture of Rathenau's life, intellectual works, and political achievements and failings.

28. *Gesammelte Schriften*, vol. 1, pp. 188–89, quoted in Stern, *Einstein's Jewish World*, p. 168.

29. Marx, "The Jewish Question," in *Selected Essays*, p. 33.

30. Nietzsche, *Genealogy*, p. 38.

31. Frankfurt, "On Bullshit," pp. 117–33.

32. Ibid., p. 130.

33. Ibid., pp. 130–31.

34. Lenard, *German Physics*. Translated in Hentschel, *Physics and National Socialism*, p. 102.

35. Stern, *The Politics of Cultural Despair.*

36. See Miller, *Einstein, Picasso.*

37. Hitler, *Mein Kampf*, p. 352.

38. Johannes Stark from the April 4, 1922, edition of the *Deutsche Tageszietung*, quoted in Rowe and Schulmann, *Einstein on Politics*, p. 13.

39. Israel, *Hundert Autoren gegen Einstein.*

40. Quoted in Hentschel, *Physics and National Socialism*, p. 1.

41. *Collected Works of Plato*, p. 402.

42. Quoted in Hentschel, *Physics and National Socialism*, p. 1.

43. Ibid., p. 2.

44. Ibid., p. 4.

45. Ibid.

46. Beyerchen, *Scientists under Hitler*, p. 88.

47. From Felix Ehrenhaft's unpublished, "My Experiences with Einstein," quoted in Clark, *Einstein*, p. 264, and Beyerchen, *Scientists under Hitler*, p. 90.

48. Beyerchen, *Scientists under Hitler*, p. 124.

49. Walker, *Nazi Science*, p. 10.

50. Beyerchen, *Scientists under Hitler*, p. 104.

51. Ibid.

52. Walker, *Nazi Science*, p.6.

53. Kant, "Physical Geography."

54. Herder, *Reflection on the Philosophy of the History of Mankind.*

55. Goethe, "The Metamorphosis of Plants."

56. Mason, *A History of the Sciences.*

57. Ibid.

58. Recapitulation theory has since been rejected for empirical reasons, but, as a working hypothesis in the nineteenth century, it was not only fruitful but also sustained a research program that produced contemporary embryology.

59. Advances in embryology also provided Nazi race theorists with what they thought to be the scientific foundations of their worldview. See Gasman, *The Scientific Origins of National Socialism.*

60. Stark, "Philipp Lenard: An Aryan Scientist," translated in Hentschel, *Physics and National Socialism*, pp. 109–10.

61. "Comment on W. Heisenberg's Reply," in *Völkischer Beobachter*, no. 59, Feb. 29, 1936, in Hentschel, *Physics and National Socialism*, p. 126.

62. Feuer, *Einstein and the Generation of Science*, p. 168.

63. Forman, "Weimar Culture, Causality, and Quantum Theory, 1918–1927."

64. Lenard, foreword to *German Physics* in Hentschel, *Physics and National Socialism*, p. 101.

65. Paul Forman, in his article "Scientific Internationalism and the Weimar Physicists: the Ideology and Its Manipulation in Germany after World War I," points out an inconsistency in this view. While there is little doubt that culture has an influence on the way a scientist approaches science, the fact is that Aryan scientists did

not see non-Aryan as nonsense. If they did then their priority disputes would have been impossible—Lenard's with Thomson and Stark's with Einstein. If these non-German scientists were not doing the same science, then how could they be accused of stealing credit for what the Aryans also did? Further, the cries for greater international recognition imply an implicit acceptance, even among the most radical proponents of scientific nationalism, of the principle of scientific internationalism.

66. Lenard, foreword to *German Physics*, pp. 100–101.

67. Lenard, *Great Men of Science*.

68. Ibid., p. 371.

69. "Comment on W. Heisenberg's Reply," in Hentschel, *Physics and National Socialism*, p. 126.

70. Lenard, *Great Men of Science*, p. 359.

71. Beyerchen, *Scientists under Hitler*, p. 124.

72. Lenard, *Great Men of Science*, p. 374.

73. Ibid., p. 375.

74. Ibid.

75. Ibid., p. 382.

76. *Völkischen Beobachter*, Jan. 3, 1921.

77. Menzel, "German Physics and Jewish Physics," in *Volkischer Beobachter*, vol. 49, no. 29, Jan. 29, 1936, translated in Hentschel, *Physics and National Socialism*, p. 119.

78. Rosenberg, *Die Spur des Juden im Wandel der Zeiten*, translated in Pois, *Alfred Rosenberg*, pp. 187–88.

79. Kraus, "Fiktion und Hypothese in der Einsteinschen Bewegunglehre," *Annalen der Philosophie*, vol. 2, no. 3 (1921), p. 359. Quoted in Reichenbach, "Present State of the Discussion on Relativity," p. 4.

80. Quoted in Frank, *Einstein*, p. 206, italics are mine.

81. Beyerchen, *Scientists under Hitler*, p. 90.

82. Born's autobiography, *My Life: Reflections of a Nobel Laureate* is a wonderful read, as is Nancy Thorndike Greenspan's biography, *The End of the Certain World: The Life and Science of Max Born*.

83. Einstein, "On the Electrodynamics of Moving Bodies," in *The Principle of Relativity*, p. 65.

Chapter Five. Did Relativity Influence the Jewish Intelligentsia?

1. See Brenner, *The Renaissance of Jewish Culture in Weimar Germany*; Neusener, *Death and Birth of Judaism*; and Mendes-Flohr, "New Trends in Jewish Thought."

2. Tönnies, *Community and Society*.

3. Brenner, *Renaissance of Jewish Culture*, p. 6.

4. Philipson, *Reform Movement in Judaism*, p. 52.

5. Quoted in Philipson, *Reform Movement in Judaism*, p. 55.

6. Ibid., p. 53.

7. Lowenstein, "Religious Life," p. 102.

8. Neusener, *Death and Birth*, p. 149.

9. For an account of Heschel's life, see Kaplan, *Spiritual Radical*.

10. "Answer to Einstein" originally published in *Aufbau*, vol. 6, no. 38, Sept. 20, 1940, p. 3, translated by Susannah Buschmeyer and published as an appendix to Kaplan, "Abraham Heschel's First American Controversies," pp. 39, 41.

11. Biographical and intellectual histories of Rosenzweig can be found in Glatzer, *Franz Rosenzweig*.

12. Rosenzweig, "Towards a Renaissance of Jewish Learning," in Glatzer, ed., *On Jewish Learning*, p. 58.

13. Brenner, *Renaissance of Jewish Culture*, pp. 79–81.

14. Ibid., pp. 26–28.

15. Friedman, *Martin Buber's Life and Work*, p. 159.

16. Einstein, *Ideas and Opinions*, p. 186.

17. Buber, *Tales of the Hasidim: Book One, The Early Masters*, p. 2.

18. Einstein, *Ideas and Opinions*, pp. 185–86.

19. Buber, *I and Thou*, pp. 59–62.

20. Ahad Ha'Am, "The Jews State and the Jews' Affliction," quoted in Vital, *Zionism*, p. 26.

21. Buber, *Israel and Palestine*, xii.

22. For a detailed discussion of the development of Einstein's Zionist views, see Jerome, *Einstein on Israel and Zionism*.

23. Einstein, "How I Became a Zionist" *Collected Papers*, vol. 7, p. 234.

24. Einstein, "Our Debt to Zionism," in *Out of My Later Years*, pp. 263–64.

25. "No More Declarations," in Mendes-Flores, ed., *A Land of Two Peoples*, p. 79.

26. Friedman, *Martin Buber's Life and Work*, pp. 8–20.

27. Biographical and autobiographical sketches of Reichenbach may be found in Reichenbach and Cohen, eds., *Hans Reichenbach*.

28. Quoted in Forman's "Weimar Culture, Causality and Quantum Theory," p. 11, which provides a detailed account of the cultural landscape of the period.

29. *The Concept of Probability in the Mathematical Representation of Reality* is a translation of Reichenbach's 1915 dissertation.

30. Reichenbach, *Theory of Relativity*.

31. Reichenbach, *Philosophy of Space and Time*.

32. Reichenbach, *Axiomatization of the Theory of Relativity*.

33. For examples of such writings in the professional and popular press, see Gimbel and Walz, eds., *Defending Einstein*.

34. For a short biography of Grelling, see Luchins, "Kurt Grelling," pp. 228–81.

35. For a discussion of the relationship between Einstein and Schlick, see Howard, "Realism and Conventionalism," pp. 618–29.

36. Translated in Stadler, *The Vienna Circle*, p. 876. This volume has a wealth of documents related to the assassination and its aftermath.

37. See Einstein's response to Reichenbach's essay, "The Philosophical Significance of the Theory of Relativity," in the collection *Albert Einstein: Philosopher-Scientist*.

38. Quoted in Feuer, *Einstein and the Generation of Science*, pp. 83–84.

39. "Phenomenology," an entry Husserl wrote for the *Encyclopedia Britannica* in 1927. Reprinted in *Husserl: Shorter Works*, p. 24.

40. Paul Forman points out in "Weimar Culture, Causality, and Quantum Theory," pp. 58–59, that the notion of an intellectual crisis became a standard part of academic discourse. "For as the notion of a crisis became a cliché, it also became an entrée, a ploy to achieve instant 'relevance,' to establish a rapport between the scientist and his auditors. By applying the word 'crisis' to his own discipline the scientist has not only made contact with his audience, but has *ipso facto* shown that his field—and he himself—is 'with it,' sharing the spirit of the times." Indeed, so ubiquitous was the use of the term "crisis" that we find it in the titles of articles of both the Aryan physics leader Stark ("The Present Crisis in German Physics") and Einstein himself ("On the Present Crisis in Theoretical Physics"). There was, Forman argues, a "craving for crises" at the time.

41. Husserl, *Crisis of European Sciences*, pp. 125–26.

42. Ibid., p. 126.

43. Ibid., p. 295.

44. Heidegger, *Being and Time*, p. 32. Italics in the original.

45. Heidegger, "Question Concerning Technology," in *Martin Heidegger: Basic Writings*, p. 332.

46. Ibid.

47. Ott, *Martin Heidegger: A Political Life*, is the most exacting source for documenting the details of this period.

48. See the interview Heidegger gave to the magazine *Der Spiegel* in 1966, "Only a God Can Save Us," and his 1945 retrospective account, "Facts and Thoughts."

49. Ott, *Martin Heidegger*, p. 143.

50. Reprinted in Runes, *German Existentialism*, p. 20. Italics in the original.

51. Ott, *Martin Heidegger*, p. 207.

52. Ibid., p. 177.

53. Ibid., p. 223.

54. Ibid., p. 199.

55. Ibid., pp. 210–23.

Chapter Six. Einstein's Liberal Science?

1. www.conservapedia.com/Theory_of_relativity, Dec. 10, 2010.

2. www.conservapedia.com/Counterexamples_to_Relativity, Dec. 10, 2010.

3. www.conservapedia.com/Theory_of_relativity, Dec. 10, 2010. "Special Relativity."

4. See Holton, *Einstein's Clocks, Poincaré's Maps*, p. 297.

5. Feuer, *Einstein and the Generations of Science*, p. 101.

6. Lorentz, *The Einstein Theory of Relativity*, p. 29.

7. www.conservapedia.com, "Mass Increase."

8. Ibid., "General Relativity."

9. Ibid., "Lack of Evidence for Relativity."

10. Ibid., "Relativity."

11. Ibid., "Special Relativity."

12. Ibid., "Relativity."

13. Earman and Glymour, "Relativity and Eclipses," pp. 49–85.

14. Kennfick, "Testing Relativity from the 1919 Eclipse," pp. 37–42.

15. www.conservapedia.com/Theory_of_relativity, "Relativity."

16. Ibid., "Political Aspects of Relativity."

17. *Time Magazine*, Nov. 23, 1970.

18. www.conservapedia.com/Theory_of_relativity, "Lack of Evidence for Relativity."

19. Ibid. The Newtonian attempts to do this involved an undiscovered planet between Mercury and the sun named Vulcan. All attempts to find such a planet failed. For an account of this search, see Marschall and Maran, *Pluto Confidential*.

20. www.conservapedia.com/Theory_of_relativity, "Lack of Evidence for Relativity."

21. Ibid., "Time Dilation and Creation Science."

22. www.conservapedia.com/Counterexamples_to_Relativity.

23. www.conservapedia.com/Theory_of_relativity, "Political Aspects of Relativity."

24. Bjerknes, "Jewish Genocide of Armenian Christians," pp. 205–6.

25. Ibid., p. 184.

26. Bjerknes, "Manufacture and Sale of Saint Einstein," p. 51.

27. Ibid. p. 14.

28.See, for example, Gianetto, "Electromagnetic Conception of Nature," pp. 765–81.

29. For a nice discussion of FitzGerald's work, see the first two chapters of Hunt, *The Maxwellians*.

30. Immanuel Kant was the first one to use the phrase "Copernican revolution" as a metaphor for a major advance that changes one's worldview in referring to his own work in the preface to *The Critique of Pure Reason*.

31. Bjerknes, *Albert Einstein, the Incorrigible Plagiarist*, p. 88.

32. Ibid., p. 115.

33. Craig, *Time and the Metaphysics of Relativity*, p. 172.

34. Bondi, *Relativity and Common Sense*, p. 168.

35. Craig, *Time and the Metaphysics of Relativity*, p. 69.

36. See Craig, *God, Time, and Eternity*, chapter 5.

37. Craig, *Time and the Metaphysics of Relativity*, p. 193.

38. Osiander in the introduction to *On the Revolution of the Heavenly Spheres*, p. 1.

39. Craig, *Time and the Metaphysics of Relativity*, p. 154.

Conclusion. Einstein's Cosmopolitan Science

1. Eze, *On Reason*, xii–xiii.

2. Ibid., p. 9.

3. Ibid.

4. Harding, *Whose Science? Whose Knowledge?*, p. 149.

5. Ibid., p. 120.

6. Ibid.

BIBLIOGRAPHY

Achinstein, Peter. *Particles and Waves*. Oxford: Oxford University Press, 1991.

Augustine. *Confessions*. Garden City, NJ: Image, 1960.

Bacon, Francis. *Novum Organon*. New York: Collier, 1900.

Balkan, David. *Sigmund Freud and the Jewish Mystical Tradition*. Boston: Beacon, 1958.

Beyerchen, Alan. *Scientists under Hitler*. New Haven: Yale University Press, 1977.

Bjerknes, Christopher Jon. *Albert Einstein: The Incorrigible Plagiarist*. Downers Grove: XTX, 2002.

———. "The Jewish Genocide of Armenian Christians." *Jewish Racism*. 2007. www.jewishracism.com/Jewish_Genocide_Enlarged.pdf (Dec. 10, 2010).

———. "The Manufacture and Sale of Saint Einstein." *Jewish Racism*. 2006. www.jewishracism.com/SaintEinstein.pdf (Dec. 10, 2010).

Bondi, Hermann. *Relativity and Common Sense*. New York: Dover, 1964.

Born, Max. *My Life: Reflections of a Nobel Laureate*. New York: Scribner, 1978.

———. *Physik im Wandel meiner Zeit*. Berlin: Braunschweig, 1958.

Boyarin, Daniel *Unheroic Conduct: The Rise of Heterosexuality and the Invention of the Jewish Man*. Berkeley: University of California Press, 1997.

Bratton, Susan Power. "Luc Ferry's Critique of Deep Ecology, Nazi Nature Laws, and Environmental Anti-Semitism." *Ethics and the Environment* 4, no. 1 (1999): 3–22.

Brenner, Michael. *The Renaissance of Jewish Culture in Weimar Germany*. New Haven: Yale University Press, 1996.

Buber, Martin. *I and Thou*. New York: Charles Scribner's Sons, 1970.

———. *Israel and Palestine: The History of an Idea*. New York: Farrar, Straus, and Young, 1952.

———. *Tales of the Hasidim*. New York: Schocken, 1991.

Carter, Miranda. *George, Nicholas, and Wilhelm: Three Royal Cousins and the Road to World War I*. New York: Knopf, 2010.

Christianison, Gale. *Isaac Newton*. Oxford: Oxford University Press, 2005.

Clark, Ronald W. *Einstein: The Life and Times*. New York: World, 1971.

Clarke, Desmond. *Descartes: A Biography.* Cambridge, UK: Cambridge University Press, 2006.

Cline, Barbara Lovett. *Men Who Made a New Physics.* Chicago: University of Chicago Press, 1965.

Cohen, Shaye J. D. *The Beginnings of Jewishness: Boundaries, Varieties, Uncertanties.* Berkeley: University of California Press, 1999.

Comte, August. *The Positive Philosophy.* New York: AMS Press, 1974.

Copernicus, Nicholas. *On the Revolution of the Heavenly Spheres.* Philadelphia: Running Press, 2002.

"Counterexamples to Relativity." www.conservapedia.com/Counterexamples_to_relativity (Dec. 10, 2010).

Craig, William Lane. *God, Time, and Eternity: The Coherence of Theism.* Dordrecht, the Netherlands: Kluwer, 2001.

———. *Time and the Metaphysics of Relativity.* Dordrecht, the Netherlands: Kluwer, 2001.

Crilly, Tony. *Arthur Cayley: Mathematician Laureate of the Victorian Age.* Baltimore: Johns Hopkins University Press, 2006.

Davidson, Eugene. *The Making of Adolf Hitler: The Birth and Rise of Nazism.* New York: Macmillan, 1977.

Dawidowicz, Lucy S. *A Holocaust Reader.* New York: Behrman House, 1976.

Descartes, René. *Discourse on Method.* Indianapolis: Hackett, 1980.

———. *Principles of Philosophy.* Boston: Reidel, 1983.

Durkheim, Emile. *Suicide: A Study in Sociology.* New York: Free Press, 1951.

———. *The Rules of Sociological Method.* New York: Free Press, 1966.

Earman, John, and Clark Glymour. "Relativity and Eclipses: The British Eclipse Expeditions of 1919 and Their Predecessors." *Historical Studies in the Physical Sciences* 11, no. 1 (1980): 49–85.

Einstein, Albert. *Albert Einstein: The Human Side.* Edited by Helen Dukas and Banesh Hoffman. Princeton: Princeton University Press, 1981.

———. *Collected Papers of Albert Einstein.* Edited by Diana K. Buchwald. Princeton: Princeton University Press, 1987.

———. *Einstein's Miraculous Year: Five Papers That Changed the Face of Physics.* Edited by John Statchel. Princeton: Princeton University Press, 2005.

———. *The Expanded Quotable Einstein.* Edited by Alice Calaprice. Princeton: Princeton University Press, 2000.

———. *Ideas and Opinions.* New York: Crown, 1954.

———. "Intellectual Autobiography." In *Albert Einstein: Philosopher-Scientist,* ed. P. A. Schilpp, 1–95. Chicago: Open Court, 1949.

———. *Out of My Later Years.* Avenel, NJ: Random House, 1956.

Euclid. *The Thirteen Books of Euclid's Elements.* Cambridge, UK: Cambridge University Press, 1926.

Eze, Emmanuel Chukwudi. *On Reason: Rationality in a World of Cultural Conflict and Racism.* Durham, NC: Duke University Press, 2008.

Ferry, Luc. *The New Ecological Order.* Chicago: University of Chicago Press, 1985.

Feuer, Lewis S. *Einstein and the Generations of Science.* New York: Basic Books, 1974.
Forman, Paul. "Scientific Internationalism and the Weimar Physicists: The Ideology and Its Manipulation in Germany after World War I," *Isis* 64, (1973): 150–80.
———. "Weimar Culture, Causality, and Quantum Theory: Adaptation by German Physicists and Mathematicians to a Hostile Environment," *Historical Studies in the Physical Sciences* 3, (1971): 1–115.
Frank, Phillip. *Einstein: His Life and Times.* New York: Knopf, 1947.
Frankfurt, Harry. "On Bullshit." In *The Importance of the Things We Care About,* 117–33. Cambridge, UK: Cambridge University Press, 1988.
Freud, Sigmund. *An Autobiographical Study.* London: Hogarth Press, 1950.
———. *The Complete Introductory Lectures on Psychoanalysis.* Edited by James Strachey. New York: W.W. Norton, 1966.
———. "Notes upon a Case of Obsessional Neurosis." In *Three Case Histories.* New York: Macmillan, 1963, 1–57.
Friedman, Maurice. *Martin Buber's Life and Work.* New York: E. P. Dutton, 1983.
Fuller, Steve. *Science.* Minneapolis: University of Minnesota Press, 1997.
Galilei, Galileo. *Dialogue Concerning Two Chief World Systems.* Berkeley: University of California Press, 1953.
Gasman, Daniel. *The Scientific Origins of National Socialism: Social Darwinism in Ernst Haeckel and the German Monist League.* New York: Elsevier, 1971.
Geroch, Robert. *General Relativity: From A to B.* Chicago: University of Chicago Press, 1981.
Gianetto, Enrico. "The Electromagnetic Conception of Nature at the Root of the Special and General Relativity Theories and Its Revolutionary Meaning." *Science & Education* 18, no. 6-7 (June 2009): 765–81.
Glatzer, Nauhum. *Franz Rosenzweig: His Life and Thought.* New York: Schocken, 1953.
Gleick, James. *Isaac Newton.* New York: Pantheon, 2003.
Goethe, Johann. *The Metamorphosis of Plants.* In Agnes Arbor, *Goethe's Botany.* Waltham: Chronica Botanica, 1946.
Gohau, Gabriel. *A History of Geology.* New Brunswick: Rutgers University Press, 1990.
Goldstein, Rebecca. *Betraying Spinoza: The Renegade Jew Who Gave Us Modernity.* New York: Schocken, 2006.
Gombay, André. *Descartes.* Malden, MA: Blackwell, 2007.
Grayling, A. C. *Descartes: The Life and Times of a Genius.* New York: Walker & Co., 2006.
Greenspan, Nancy Thorndike. *The End of the Certain World: The Life and Science of Max Born.* New York: Basic, 2005.
Harding, Sandra. *Whose Science? Whose Knowledge? Thinking from Women's Lives.* Ithaca: Cornell University Press, 1991.
Hegel, Friedrich. *Phenomenology of Spirit.* New York: Oxford Univeristy Press, 1977.

Heidegger, Martin. *Being and Time*. New York: Harper and Row, 1962.
———. "Facts and Thoughts." Translated by Karsten Harries in *Review of Metaphysics* 38 (March 1985): 467–502.
———. *Martin Heidegger: Basic Writings*. Edited by David F. Krell. New York: HarperCollins, 1977.
———. "Only a God Can Save Us." Translated by Maria Alter and J. D. Caputo, *Philosophy Today*, vol. 20, 1976, pp. 267–84.
Hentschel, Klaus."Bernhard Bavink (1879–1947): der Weg eines Naturphilosophen vom deutschnationalen Sympathisanten der NS-Bewegung bis zum unbequemen Non-Konformisten." *Suddhofs Archiv*. vol. 77, no. 1 (1993): 1–32.
———. *Physics and National Socialism: An Anthology of Primary Sources*. Basel, Switzerland: Birkhauser Verlag, 1996.
Herder, Johann. *Reflection on the History of Mankind*. Chicago: University of Chicago Press, 1968.
Hibbert, Christopher. *The House of Medici: Its Rise and Fall*. New York: Morrow, 1975.
Hilberg, Raul. *The Destruction of the European Jews*. Chicago: Quadrangle Press, 1961.
Hitler, Adolf. *Mein Kampf*. Boston: Houghton-Mifflin, 1971.
Holton, Gerald. *Einstein's Clocks, Poincaré's Maps*. New York: W.W. Norton, 2003.
Howard, Don. "Realism and Conventionalism in Einstein's Philosophy: The Einstein-Schlick Correspondence." *Philosophia Naturalis* 21 (1984): 618–29.
Hume, David. *Treatise on Human Nature*. Oxford: Clarendon, 1888.
Hunt, Bruce. *The Maxwellians*. Ithaca: Cornell University Press, 1991.
Husserl, Edmund. *Husserl: Shorter Works*. Edited by Frederick Elliston and Peter McCormick. South Bend, IN: Notre Dame University Press, 1981.
———. *The Crisis of European Sciences and Transcendental Phenomenology*. Evanston, IL: Northwestern University Press, 1970.
Israel, Hans, ed. *Hundert Autoren gegen Einstein*. Innsbruck, Austria: University of Innsbruck Press, 2007.
Jammer, Max. *Einstein and Religion*. Princeton: Princeton University Press, 2000.
Jerome, Fred. *Einstein on Israel and Zionism*. New York: St. Martin's, 2009.
———. *The Einstein File: J. Edgar Hoover's Secret War against the World's Most Famous Scientist*. New York: St. Martin's, 2002.
Jerome, Fred, and Rodger Taylor. *Einstein on Race and Racism*. New Brunswick: Rutgers University Press, 2005.
Jones, Sheila. *The Quantum Ten*. Oxford: Oxford University Press, 2008.
Kant, Immanuel. *Critique of Pure Reason*. London: Bell and Daldy, 1866.
———. "Physical Geography." In *Race and the Enlightenment*, ed. Emmanuel Chudwudi Eze, 58–64. Cambridge, UK: Blackwell, 1997.
Kaplan, Edward. "Abraham Heschel's First American Controversies: Einstein, War, and the Living God." *Conservative Judaism* 54, no. 4 (2003): 26–41.
———. *Spiritual Radical: Abraham Joshua Heschel in America*. New Haven: Yale University Press, 2007.
Kennefick, Daniel. "Testing Relativity from the 1919 Eclipse: A Question of Bias." *Physics Today* 62, no. 3 (2009): 37–42.

Kitchen, Martin. *A History of Modern Germany: 1800–2000*. Malden, MA: Blackwell, 2006.

Kühl, Stefan. *The Nazi Connection: Eugenics, American Racism, and German National Socialism*. Oxford: Oxford University Press, 1994.

Kuhn, Thomas. *The Structure of Scientific Revolutions*. Chicago: University of Chicago Press, 1952.

Latour, Bruno, and Steve Woolgar. *Laboratory Life: The Social Construction of Scientific Facts*. Princeton: Princeton University Press, 1979.

Lenard, Philipp. *Great Men of Science*. New York: Macmillan, 1933.

Loeb, James, ed. *Apostolic Fathers*. Cambridge, MA: Harvard University Press, 1912.

Lorentz, H. A. *The Einstein Theory of Relativity: A Concise Statement*. New York: Brentano's, 1920.

Lorentz, H. A., Albert Einstein, Hermann Minkowski, and Hermann Weyl. *The Principle of Relativity: A Collection of Original Memoirs on the Special and General Theory of Relativity*. New York: Dover, 1952.

Lowenberg, Peter. "Sigmund Freud's Psycho-Social Identity." In *100 Years of Psychoanalysis*, ed. Andre Haynal and Ernst Falzeder, 135–50. London: H. Karnak, 1966.

Lowenstein, Steven. "Religious Life." In *German-Jewish History in Modern Times*, ed. Michael Meyers, vol. 3, 103–24. New York: Columbia University Press, 1997, vol. 3.

Luchins, Edith, and Abraham Luchins. "Kurt Grelling: Steadfast Scholar in a Time of Madness." *Gestalt Theory* 22, no. 4 (2000): 228–81.

Luther, Martin. *Martin Luther: Selections from His Writings*. Edited by John Dillinberger. Garden City: Anchor, 1961.

Mach, Ernst. *The Science of Mechanics: A Critical and Historical Account of Its Development*. Chicago: Open Court, 1960.

Manuel, Frank. *A Portrait of Isaac Newton*. Cambridge, MA: Belknap, 1968.

Marschall, Laurence, and Stephen Maran. *Pluto Confidential*. Dallas: BenBella, 2009.

Marty, Martin. *Martin Luther: A Life*. New York: Penguin, 2004.

Marx, Karl. *Capital*. New York: Penguin, 1980.

———. "The Jewish Question." In *Karl Marx: Selected Essays*, ed. H. J. Stenning, 16–40. Middlesex: Echo, 2008.

Mason, Stephen. *A History of the Sciences*. New York: Dover, 1952.

Mehra, Jagdish. *Einstein, Hilbert, and the Theory of Gravitation: Historical Origins of the General Relativity Theory*. Boston: D. Reidel, 1974.

Mendes-Flores, Paul, ed. *A Land of Two Peoples*. Chicago: University of Chicago Press, 1983.

———. *New Trends in Jewish Thought*. Vol. 3 of *German-Jewish History in Modern Times*, ed. Michael Meyers. New York: Columbia University Press, 1997.

Merton, Robert K. *Science, Technology, and Society in 17th Century England*. New York: H. Fertig, 1970.

Meyers, Eric M. "Jewish Culture in Greco-Roman Palestine." In *Cultures of the Jews: A New History*, ed. David Biale, 138–79. New York: Schocken, 2002.
Miller, Arthur I. *Albert Einstein's Special Theory of Relativity: Emergence (1905) and Early Interpretation (1905–1911)*. Reading, PA: Addison-Wesley, 1981.
———. *Einstein, Picasso*. New York: Basic, 2001.
Mitteraur, Michael. *A History of Youth*. Cambridge: Blackwell, 1993.
Mommsen, Hans. *From Weimar to Auschwitz*. Princeton: Princeton University Press, 1991.
———. *The Rise and Fall of Weimar Democracy*. Chapel Hill: University of North Carolina Press, 1996.
Neusener, Jacob. *Death and Birth of Judaism*. New York: Basic, 1987.
Newton, Isaac. *Isaac Newton's Papers and Letters on Natural Philosophy*. Edited by I. B. Cohen. Cambridge: Harvard University Press, 1958.
———. *Mathematical Principles of Natural Philosophy*. Berkeley: University of California Press, 1974.
———. *Opticks*. New York: Dover, 1952.
Nietzsche, Friedrich. *On the Genealogy of Morals*. New York: Random House, 1967.
North, John. *Cosmos*. Chicago: University of Chicago Press, 2008.
Ott, Hugo. *Martin Heidegger: A Political Life*. New York: Basic, 1993.
Pais, Abraham. *Subtle Is the Lord: The Science and the Life of Albert Einstein*. Oxford: Oxford University Press, 1982.
Parshall, Karen. *James Joseph Sylvester: A Jewish Mathematician in a Victorian World*. Baltimore: Johns Hopkins University Press, 2006.
———. *James Joseph Sylvester: Life and Work in Letters*. Oxford: Clarendon, 1998.
Philipson, David. *The Reform Movement in Judaism*. New York: Macmillan, 1931.
Plato. *Apology*. In *The Collected Works of Plato*, 401–26. New York: Random House, 1920.
———. *Republic*. New York: Anchor, 1973.
Poincaré, Henri. *Science and Hypothesis*. New York: Dover, 1952.
———. *The Value of Science*. New York: Dover, 1958.
Proctor, Robert. *Racial Hygiene: Medicine under the Nazis*. Cambridge, MA: Harvard University Press, 1988.
Pulzer, Peter. *Jews and the German State*. Cambridge, UK: Blackwell, 1992.
Raman, V. V. "Relativity in the Early Twenties: Many-Sided Reactions to a Great Theory," *Indian Journal of History of Science* 7 (1972), 119–45.
Reichenbach, Hans. *Axiomatization of the Theory of Relativity*. Berkeley: University of California Press, 1965.
———. *The Concept of Probability in the Mathematical Representation of Reality*. Chicago: Open Court, 2008.
———. *Defending Einstein: Hans Reichenbach's Early Writings on Space, Time, and Motion*. Edited by Steven Gimbel and Anke Walz. Cambridge, UK: Cambridge University Press, 2006.
———. *Hans Reichenbach: Selected Writings, 1909–1953*. Edited by Maria Reichenbach and Robert S. Cohen. Boston: D. Reidel, 1978.

———. *Modern Philosophy of Science.* London: Routledge and Kegan Paul, 1959.
———. *Philosophy of Space and Time.* New York: Dover, 1958.
———. *Theory of Relativity and A Priori Knowledge.* Berkeley: University of California Press, 1965.
Reid, Constance. *Hilbert.* New York: Springer, 1970.
Renn, Jürgen, and Robert Schulmann. *Albert Einstein / Mileva Maric: The Love Letters.* Princeton: Princeton University Press, 1992.
Reulecke, Jürgen. "The Battle for the Young: Mobilising Young People in Wilhelmine Germany." In *Generations in Conflict: Youth Revolt and Generation Formation in Germany, 1770–1968.* Edited by M. Roseman. Cambridge, UK: Cambridge University Press, 1995, 92–104.
Rife, Patricia. *Lise Meitner and the Dawn of the Nuclear Age.* Boston: Birkhauser, 1999.
Rosenberg, Alfred. *Alfred Rosenberg: Selected Writings.* Edited by Robert Pois. London: Jonathan Cape, 1970.
Rosenzweig, Franz. *On Jewish Learning.* Edited by Nauhum Glatzer. Madison: University of Wisconsin Press, 1955.
Rowe, David E. "'Jewish Mathematics' at Göttingen in the Era of Felix Klein." *Isis* 77, no. 3 (Sept. 1986): 422–49.
Rowe, David E., and Robert Schulmann. *Einstein on Politics.* Princeton: Princeton University Press, 2007.
Rubenstein, Richard. *When Jesus Became God: The Struggling to Define Christianity during the Last Days of Rome.* New York: Mariner, 2010.
Runes, Dagobert. *German Existentialism.* New York: Wisdom Library, 1965.
Satlow, Michael. "Defining Judaism: Accounting for 'Religions' in the Study of Religion." *Journal of the American Academy of Religion* 74, no. 4 (2006): 837–60.
———. "'Wasted Seed': The History of a Rabbinic Idea." *Hebrew Union College Annual* 65 (1994): 137–75.
Schopenhauer, Arthur. *On Human Nature: Essays in Ethics and Politics.* New York: Macmillan, 1910.
Sime, Ruth Lewin. *Lise Meitner: A Life in Physics.* Berkeley: University of California Press, 1996.
Sobel, Dava. *Galileo's Daughter.* New York: Walker & Company, 1999.
Spinoza, Baruch. *Ethics.* New York: Penguin, 1996.
Stadler, Friedrich. *The Vienna Circle: Studies in the Origins, Development, and Influence of Logical Empiricism.* New York: Springer, 1997.
Steinsaltz, Adin, ed. *The Essential Talmud.* New York: Basic, 1976.
Stern, Fritz. *Einstein's Jewish World.* Princeton: Princeton University Press, 2001.
———. *The Politics of Cultural Despair: A Study in the Rise of Germanic Ideology.* Berkeley: University of California Press, 1961.
Sylvester, J. J. "Note on a Proposed Addition to the Vocabulary of Ordinary Arithmetic." *Nature*, 1888, vol. 47, issue 946: 152–53.
"Theory of Relativity". www.conservapedia.com/Theory_of_relativity (Dec. 10, 2010).

Time. "A Victory for Relativity." Nov. 23, 1970.

Tönnies, Ferdinand. *Community and Society*. East Lansing: Michigan State University Press, 1963.

Veblen, Thorstein. "The Intellectual Pre-eminence of Jews in Modern Europe." *Political Science Quarterly* 34, no. 1 (1919): 33–42.

Viereck, George Sylvester. "Freud's First Interview on Psycho-Analysis." *New York American*, Aug. 19, 1923.

Vital, David. *Zionism: The Formative Years*. New York: Clarendon, 1982.

Wagner, Richard. "Judaism in Music," in *Richard Wagner's Prose Works*, vol. 3, ed. William Ashton Ellis, 79–100. London: William Reeves, 1907.

Walker, Mark. *Nazi Science: Myth, Truth, and the German Atomic Bomb*. New York: Plenum, 1995.

Weikart, Richard. *From Darwin to Hitler: Evolutionary Ethics, Eugenics, and Racism in Germany*. New York: Palgrave, 2004.

Weindling, Paul. *Health, Race and German Politics between National Unification and Nazism, 1870–1945*. Cambridge, UK: Cambridge University Press, 1989.

Weiner, Marc. *Richard Wagner and the Anti-Semitic Imagination*. Lincoln: University of Nebraska Press, 1995.

Westphal, Richard. *Never at Rest: A Biography of Isaac Newton*. Cambridge, UK: Cambridge University Press, 1980.

White, Michael. *Isaac Newton: The Last Sorceror*. Reading, PA: Addison-Wesley, 1997.

Wohl, Robert. *The Generation of 1914*. Cambridge, MA: Harvard University Press, 1979.

Yaglom, I. M. *Felix Klein and Sophus Lie: Evolution of the Idea of Symmetry in the Nineteenth Century*. Boston: Birkhauser, 1988.

INDEX